E-Governance in India

Sunil K. Muttoo • Rajan Gupta • Saibal K. Pal

E-Governance in India

The Progress Status

Sunil K. Muttoo
Department of Computer Science,
Faculty of Mathematical Sciences
University of Delhi
Delhi, Delhi, India

Rajan Gupta
Deen Dayal Upadhyaya College
University of Delhi
Delhi, Delhi, India

Saibal K. Pal
Defence Research and Development
Organisation
Delhi, Delhi, India

ISBN 978-981-13-8851-4   ISBN 978-981-13-8852-1   (eBook)
https://doi.org/10.1007/978-981-13-8852-1

Cover design: eStudio Calamar

This Palgrave Macmillan imprint is published by the registered company Springer Nature
Singapore Pte Ltd.
The registered company address is: 152 Beach Road, #21-01/04 Gateway East, Singapore
189721, Singapore

Dedicated to
*My "**Gurumaa**" for holding my hand in Life,*
*My "**Parents**" for making me stand in Life,*
*My "**Brother**" for helping me progress in Life, and*
*My "**Wife**" for supporting and loving me unconditionally in Life.*

Rajan Gupta

PREFACE

This book on E-Governance in India has got three major things on offer—introduction to E-Governance, status of E-Governance in India through various measures, and finally the progress of E-Governance initiatives in India through various case studies.

Firstly, the book discusses about the concepts around E-Governance in general. What are the various principles of Good Governance and how do they lead to the principles of E-Governance, are the topics discussed in Chap 1. Then the objectives of the E-Governance are discussed from the understanding of a naïve user. The historical development of E-Governance around the world and its rise in few developed nations like USA, UK, New Zealand, has been discussed. Also, the United Nation's survey on E-Governance development is discussed about. The detailing of the concepts of the E-Governance will give background to the readers of the book, so that they can have understanding about the progressive metrics discussed in the subsequent chapters.

Secondly, the book discusses about the establishment of E-Governance in India in detail and then compares the progress in Indian states through different measures and metrics. The structure of the E-Governance in India has been explained in Chap. 2, which also includes the explanation of the details related to National E-Governance plan (NeGP). For any reader of the book, it is important to understand the E-Governance set-up in India and how differently the various components have been implemented in this country. This allows the user to assess the breadth and depth of infrastructure related to E-Governance in a developing country like India. Then, through Chap. 3, the status and progress of India at the world level has been analyzed. This shows the performance of the execution of NeGP in India and also shares where India is lacking with respect to other countries, which are performing well on the implementation front of E-Governance. Then, the regional level and sector level analysis and comparison for various Indian states and union territories have been conducted in Chap. 4. This will help in understanding the progress of E-Governance in various states and sectors of India. The weak and strong regions are indicated in this chapter through different types of data curated from other reliable resources. This offering point of the book gives a comprehensive view of E-Governance performance and progress in India.

Thirdly, the progress of E-Governance in India through case studies has been carried out in Chaps. 5 and 6. Initially, Chap. 5 discussed the scope of improvement of E-Governance in India through futuristic trends emerging from the SMAC concept, that is, social, mobile, analytics and cloud. How E-Governance will use these four concepts, for their betterment, is discussed in Chap. 5. The various shortcomings and gaps for the current execution, which can be covered up using technology know-how of SMAC are also discussed. And lastly, Chap. 6 presents the various digital and E-Governance initiatives taken by Government of India. The concept, performance and measures related to the various E-Governance initiatives are presented up to the year 2017. The reader will get an idea about the various current and future initiatives of E-Governance in India.

The current book is a combination of theoretical and practical concepts defined over various aspects of E-Governance in India. This book can serve as the first-stage book for any individual working in Indian Territory on E-Governance. Major concepts and progress will help the individuals to work ahead of the basics and develop a more useful framework. This book will also be useful for government departments and employees for tracking the status of E-Governance in India. The government officials can use the data and information to make roadmaps for the improvement of E-Governance in their respective domains. Also, the academicians will get a good base prepared for their research work to be carried out in the field of E-Governance in India. Since, basic concepts, datasets and information about India has been compiled in this book, therefore, it will serve well to the students community as well, who are studying subjects like politics, IT, information systems, law, journalism, and computer science.

This book offers multi-disciplinary inputs to the people working in different fields toward E-Governance. The analysis of data has been presented through graphs and tables for better understanding of the reader.

Delhi, Delhi, India	Sunil K. Muttoo
Delhi, Delhi, India	Rajan Gupta
Delhi, Delhi, India	Saibal K. Pal

Disclaimer

The purpose of this book is to provide information on the subject of "E-Governance" in the Indian region compiled from various sources. The book does not intend to provide any political stand nor is it related to any political organization in India or abroad. The book has been written from a neutral viewpoint and does not represent views of any profitable organization.

The data has been collected from various secondary sources available in the form of public reports, websites, mobile applications, newspapers, internet blogs, published books, research papers and journal papers. Due referencing has been given to all the sources as per the best of the knowledge of the authors.

Content and data in the field of E-Governance is very dynamic and may change during the publishing of the book. Therefore, a reference point of year 2017 has been taken while presenting the facts and figures in the book. In case readers wish to consider updated statistics, the various websites and other resources given at the end of this book may be visited again.

The information given in this book is considered true as per the data and content given on the various reference links given at the end. So, the details should be considered as indicative based on the already available information at various secondary sources. For accurate and updated information, readers are advised to visit the concerned department, weblink, or any other source of information, and collect the required data and information. The authors are not responsible for any information not correctly presented on the various secondary sources given in the reference list.

ACKNOWLEDGMENTS

The authors of this book would like to gratefully and sincerely thank all the people who have supported them during the journey of writing this book, to only some of whom it is possible to mention here.

The authors would like to thank current and former faculty members and Head of Department of Computer Science, University of Delhi—Dr. Vasudha Bhatnagar, Dr. Punam Bedi, Dr. Naveen Kumar, Mr. P. K. Hazra, Dr. Neelima Gupta, and Ms. Vidya Kulkarni, for helping in providing infrastructure and resources related to this book. The authors also thank teaching and non-teaching staff members for their support in the Department of Computer Science, University of Delhi.

Also, the authors would like to thank the team of Deen Dayal Upadhyaya College, University of Delhi, under the guidance of Dr. S. K. Garg and Dr. Hemchand Jain for providing their support toward the writing of this book.

Authors would like to thank the authorities of Defense Research & Development Organization (DRDO), Government of India, for providing valuable inputs toward the research work carried out for this book.

Finally, this work would not have been possible without invaluable support from the publishing team of Palgrave Macmillan. This book also recognizes incredible support from the book's endorsers, and the authors' guru, mentors, family and friends. So the authors would like to thank them all from bottom of their heart.

The authors have collectively worked on the doctoral thesis titled "Techniques for Improvement of E-Governance in Developing Nations" submitted at University of Delhi and would like to acknowledge that the idea of working on this book was derived from that thesis.

CONTENTS

About the Authors

Sunil K. Muttoo is Professor at the Department of Computer Science, Faculty of Mathematical Sciences, University of Delhi, India. He received his PhD in Coding Theory and M.Phil. in Mathematics from the University of Delhi and M.Tech. in Computer Science and Data Processing from IIT, Kharagpur. His areas of interest include Information Hiding, Coding Theory and E-Governance. He has over 100 publications at national and international forums.

Rajan Gupta is a Research and Analytics Professional and is the Corresponding Author of the current book. He has done his PhD (Information Systems) in the area of E-Governance from the Department of Computer Science, University of Delhi. He completed his Master in Computer Application (MCA) from University of Delhi, Post-graduate Program in Management (PGPM) from IMT Ghaziabad, and Executive Program in Business Intelligence and Analytics (EPBABI) from IIM, Ranchi. He is NET-JRF qualified under University Grant Commission, India, and holds a certificate in Consulting from Consultancy Development Centre (CDC), DSIR, Ministry of Science and Technology, Government of India. He is one of the few Certified Analytics Professionals (CAP) around the world and is currently serving as CAP Ambassador in Asia Region (INFORMS, USA). He has been accredited with GStat from American Statistical Association. He has worked with firms like Samsung and Inauctus [TCF Consultancy] in their research and analytics department. He has also been associated as Assistant Professor on Ad-hoc basis with Deen Dayal Upadhyaya College, University of Delhi, Department of

Computer Science (Faculty of Mathematical Science)—University of Delhi, and IMT—Ghaziabad for delivering lectures in Computer Science, IT and Management. His area of interest includes E-Governance, Public Information Systems, Multimedia Data processing, Data Mining and Data Analytics. He has over 50 publications at various national and international forums.

Saibal K. Pal is Director of Cyber Security at Defense Research & Development Organization (DRDO), Delhi. He has worked as Senior Scientist with Scientific Analysis Group (SAG), DRDO for many years and has been awarded "Scientist of the Year" from Government of India. He received his PhD in Computer Science from University of Delhi and is an Invited Faculty & Research Guide at a number of national institutions. His areas of interest are Information & Network Security, Computational Intelligence, Information Systems and Electronic Governance. He has more than 100 publications in books, journals and international conference proceedings.

LIST OF FIGURES

List of Tables

CHAPTER 1

Introduction

Sunil K. Muttoo

1.1 Meaning of Governance

The simple definition of Governance is the process of decision making and implementation of those decisions. The concept of governance is as old as human civilization. It can take various forms as corporate governance, international governance, and local or national governance. The process of governance involves formal and informal actors and structures to arrive at and implement a decision. Government is one of the main actors in the process of governance. There are other actors involved such as in rural areas NGOs, leaders, financial institutions and political parties participate in governance process and in urban areas media, lobbyists, multinational corporations and international donors influence decision-making process (United Nations 2016).

Governance includes all the formal and informal processes used by the government. The actors are involved in the process of interaction and decision making. Governance can be related to different entities and in different manners. It can be associated with a particular 'level' of governance in a type of organization (public governance, corporate governance, non-profit governance, global governance and project governance), a particular 'field' of governance with specific outcome (environmental governance, IT governance and internet governance) and a particular model of governance related to a theory (regulatory governance, participatory governance, multilevel governance and collaborative governance).

© The Author(s) 2019
S. K. Muttoo et al., *E-Governance in India*,
https://doi.org/10.1007/978-981-13-8852-1_1

OECD defines Public Governance as "*the formal and informal arrangements that determine how public decisions are made and how public actions are carried out, from the perspective of maintaining a country's constitutional values in the face of changing problems, actors and environments*" (Bouckaert 2006). It is the economic, political and administrative authority to manage the governmental activities.

Public governance, in general, occurs in three broad ways.

(a) Through public-private partnership (PPP) networks or with the collaboration of community organizations;
(b) Through market mechanisms where market principles of competition are used to allocate resources while in service under government regulation; and
(c) Using top-down methods that engage governments and bureaucracy of the State.

1.2   Good Governance

Good Governance refers to the concept of practicing various processes and the participation of Governance in a fair manner. There are various characteristics associated with Good Governance like transparency, efficiency, accountability, effectiveness, participation, responsiveness, consensus oriented, and equitability. The major advantages associated with good governance are reduction in corruption, equality in participation of various decision makers irrespective of caste, creed or sex, and treatment of everyone in equal manner. Good governance is responsive to the needs and requirements of the society (Gisselquist 2012).

Good governance encompasses following *rule of the law, effective participation, political pluralism, transparent and accountable processes, efficient public sector, information and education, access to knowledge, sustainability, equity, empowerment of people, attitude, values* and *responsibility*. Good governance relates to political and institutional processes to achieve developmental goals. Good governance involves handling public affairs, manage public resources, and follow rule of law. Good governance is the extent of civil, cultural, economic, political and social rights given to the citizens. The Commission on Human Rights has highlighted key characteristics of good governance (United Nations Human Rights 2016): Transparency, Responsibility, Responsiveness, Participation and Accountability.

Criteria of goodness have been explained by United Nations (2016) as follows.

1. Good governance lacks parsimony. Good governance has multiple definitions and there is a need to understand the different aspects.
2. Good governance also lacks differentiation. Many well-governed countries have similar governance structure and there is little difference in their governance methods.
3. Good governance lacks coherence. It has many possible characteristics that generally do not belong together such as respect for human rights and efficient banking regulations.
4. Good governance also lacks theoretical utility. The theory formulation and the related project of hypothesis testing are confusing, but analysts can easily define it in the best possible manner.

1.3   PRINCIPLES OF GOOD GOVERNANCE

Good governance in the public sector aims to ensure the entities act in the public interest. It requires having a strong commitment to integrity, ethical values, rule of law and comprehensive stakeholder engagement. Good governance, apart from acting in the public interest, also requires defining outcomes in terms of sustainable economic, social, and environmental benefits. It aims to determine the interventions necessary for the achievement of intended outcomes and helps in developing the capacity of the entity, that is, leadership, capability of managing risks and performance through internal control. Good governance also requires implementation of good practices that allow transparency and promote accountability (CIFPA 2013).

UNDP Governance and Sustainable Human Development has defined certain principles of good governance in 1997. These principles have universal recognition and they are based on related UNDP text. The five good governance principles stated by UNDP are legitimacy and voice, direction, performance, accountability and fairness (Graham et al. 2003).

1.3.1   Legitimacy and Voice

The legitimacy and voice principle are based on the text that all men and women should get equal chance to participate in decision making whether it is direct or through legitimate intermediate institute representing their interests. This principle is based on the concept of freedom of speech and capacities to participate beneficially. The aim is to reach the best decision with the consensus of the group on the policies and procedures.

1.3.2   Direction

Good governance involves a broad perspective of leaders and public on governance. Further, human development and a sense of what is required for such development constitute it as well. It is the understanding of historical, cultural and social complexities to reach at the best decision.

1.3.3   Performance

Good governance requires responsiveness of the entities, an ability of the institutions and processes to try to serve the needs of all stakeholders. It involves achieving effectiveness and efficiency in meeting the needs while making the best of available resources.

1.3.4   Accountability

The principle of accountability and transparency go hand in hand. This principle holds that the decision makers in government, the private sector and civil society organizations are accountable to the public and to the institutional stakeholders. This accountability depends on the organizations and the type of decisions.

Transparency allows a free flow of information. The information must be directly accessible to those concerned with them to understand and monitor them based on which decisions are taken.

1.3.5   Fairness

The principle of fairness includes equity and rule of law for everyone. The principle of equity states that all men and women have equal opportunities to improve or maintain their well-being. The Rule of Law states that the legal frameworks should be fair and must be impartial especially in case of laws on human rights.

1.4 E-GOVERNANCE

E-Governance has been introduced in India via usage of technology and implementing the usage of computers within the various departments of Government of India. This was done to enhance the concept of "Minimum Government, Maximum Governance". It involves inclination of government toward service orientation, citizen centricity, and the transparency of governance. Prior initiatives of E-Governance help in preparing the way ahead and different progressive strategies for the same. For this, a structured approach has been implemented to speed up the process at various levels of National and State (regional) importance, which are bound by a common strategy and vision. E-Governance helps in cost saving through sharing of core and support infrastructure. It also enables interoperability through standards and enables strong interaction between the citizens of the country and government.

The government introduced the National E-Governance Plan (NeGP) to overview the governance initiatives in India with the services spread across different states and Union Territories (UTs) with a collective vision of good governance. The implementation of technology-based solutions, creating infrastructure, providing access to the internet and using digitization to various existing data records, are all included under the plan of NeGP. The major objective communicated through the NeGP's Vision Statement is to reduce the gap between the public services and their access to the citizens. The vision statement communicates that the initiative aims to make government services to the citizens through reliable service delivery outlets. And to ensure transparency and efficiency of such services, citizens can avail those services at a reliable cost (DeitY 2016a, b, c).

1.5 E-GOVERNANCE AND E-GOVERNMENT

E-Governance, as represented by various researchers in the past, is a more exhaustive concept than E-Government, which makes use of ICT by the government and the society. Aim of E-Governance is to increase citizen's participation rate in the government services and institutions. It also involves usage of Information and Communication Technology by various government officials and politicians for publicity, information passing or to receive feedback (Howard 2001; Bannister and Walsh 2002).

The major difference in E-Government and E-Governance lies in their purpose. E-Government's focus is on the stakeholders outside the

Table 1.1   Framework for E-Government versus E-Governance

		Focus	
		Outside	Inside
Type of organization	Public sector-government agency	E-Government (extranet and internet)	E-Governance (intranet)
	Private sector	Inter-organizational systems (extranet and intranet)	E-Governance (intranet)

Palvia and Sharma (2007)

organization such as the government or public sector at the city, state, national or international level, whereas E-Governance has its focus on the management of internal issues, that is, within the organization, whether it is a public or a private organization. So, E-Government is said to be concerned with utilization of information for managing organizational issues while E-Government involves interaction of governmental agency with outside constituencies (Table 1.1).

E-Governance is concerned with internally focused utilization of information technologies and internet to manage human, material and capital resources of the organization. It takes into account activities of government employees on the internet such as calculation of retirement benefits, access to applications, and collaboration with other government employees anytime. E-Government, on the other hand, involves interaction of the government agency with outside constituencies. The interaction can be with citizens, businesses or with other governmental agencies. It involves government agencies deploying information and internet technologies to fulfill their responsibilities of collecting taxes and using revenues for defense, education, security and health care (Palvia and Sharma 2007).

1.6   OBJECTIVES OF E-GOVERNANCE

The governance initiatives have the following aims (Ganore 2011).

1.6.1   To Build an Informed Society

The governance initiatives have the objective of building an empowered society. The governance initiatives of the government aim to inform people and provide access of information about the Government to the public.

1.6.2   To Increase Interaction between Government and Citizens

The Government and Citizens generally do not interact. The government does not receive much feedback from the citizens. The governance aims at interacting with people, build a feedback framework, and to make a system where people can make the Government aware of their problems.

1.6.3   To Encourage Citizen Participation

The governance helps in fulfilling the true purpose of democracy that requires participation of each individual citizen. The governance procedure helps in enhancing the participation rate of the public by improving the feedback, access to information and overall participation of the citizens in the decision-making process of the government.

1.6.4   To Bring Transparency in the Governing Process

The governance process also carries the objective of making the Governing process transparent by disclosing all the Government data and information to the people for access. It enhances people's knowledge about the decisions, and policies of the Government.

1.6.5   To Make the Government Accountable

Government has the responsibility of every act and decision taken by it. The governance process helps make the Government more accountable by bringing transparency and making the citizens more aware of government's decisions.

1.6.6   To Reduce the Cost of Governance

Another major aim of E-Governance is to reduce cost of governance by reducing the expenditure on physical delivery of information and services. Reducing the use of stationery, which amounts to most of the government's expenditure, can help in cutting down the cost. Also, it reduces the need for physical communication which reduces the time and cost of the government.

1.6.7   *To Reduce the Reaction Time of the Government*

The government generally delays in solving public problems and even does not reply to people's queries in an efficient manner. This kind of behavior is usually evident due to red-tapism and other reasons. Thus, E-Governance procedure helps the Government to react and reduce people's queries and problems.

1.7   SEGMENTS OF E-GOVERNANCE

E-Governance services are shared with three target groups *namely*, citizens, businesses and other government departments. The external objectives focus on citizens and businesses while internal objectives are concerned with other governmental constituencies. The four models of E-Governance as identified by Yadav and Singh (2013) are as follows (Fig. 1.1).

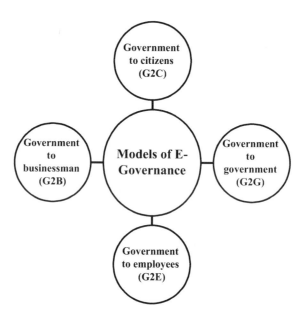

Fig. 1.1   Models of E-Governance. (Yadav and Singh 2013)

1.7.1   *Government to Citizens (G2C)*

This model describes the government services and information shared with the citizens. It provides access to the citizens to the government websites and applications. It facilitates citizens' interaction with the government through payment of online bills, online registration of applications, tax payments, online filling of complaints, accessing any kind of online information and renewal of license.

1.7.2   *Government to Government (G2G)*

This model involves interaction, sharing of data and electronic exchanges amongst governmental departments. Inter and intra-departmental exchanges at National, State and Local levels are carried on through activities like preparation, approval, distribution, and storage of all governmental documents.

1.7.3   *Government to Employees (G2E)*

This model describes the transparency between the government and its employees. This can be maintained through internal communications like data submission, complaints, leave applications, internal guidelines and regulations.

1.7.4   *Government to Businessman (G2B)*

This model describes the relationship between the government and the private sector businesses. It majorly involves dealing with the suppliers through collection of taxes, complaints or dissatisfaction, payment of bills and sharing of information and data.

1.8   RISE OF E-GOVERNANCE

E-Governance rose in government department while E-Commerce gained momentum in private companies. E-Governance is an initiative to make the government departments more efficient. The initial efforts of E-Governance were more irregular on part of IT professionals in government. Initially, it made its presence felt through a limited scale and coverage. The arrival of World Wide Web (WWW) made it feasible for the government departments to introduce their websites and portal. The rise of E-Governance includes a series of evolution levels in concept and applications of IT in government. The growth of E-Governance in India is not same as it did in USA or Europe, but it has now become mandatory in government departments, and is not merely an option for them.

The early phase of E-Governance is mostly related to problems in Information Resource Management (IRM), Strategic Information Management (SIM), performance yardsticks to measure practices. The government departments were initially bothered about the broad strategic issues related to information management. The second phase is related to the rise of the World Wide Web and the problems connected to the management of WWW, design of websites, assessment of websites, and so on. Then next phase of E-Governance was about problems relating to security, accountability, reliability, legislations, participation tools, and organization (Mitra 2012).

The increasing importance of electronics made the government of India to establish a Department of Electronics in 1970. The establishment of National Informatics Centre (NIC) in 1977 was another step taken towards E-Governance which brought information and communication in limelight. However, E-Governance in India actually gained momentum with the launch of *NICNET* in 1987 satellite-based computer network nationwide. The government of India also launched District Information System of the National Informatics Centre (DISNIC) with the aim to computerize all the district level offices of the country. Under this program, State governments were offered free hardware and software. NICNET was then extended to all district headquarters via the State capitals by 1990 (Mitra 2012).

The first E-Governance project in India was started in Kerala by the name of AKSHAYA and it focused on E-Learning, E-Transaction, E-Governance, information and communication and provides social and economic power to the public (Mitra 2012).

E-Governance has been launched worldwide at different points of time. The governments around the world have taken important steps for the proper implementation and expansion of E-Governance and made working of government more efficient, responsive and transparent in the respective countries. Countries like UK, USA, New Zealand, and Brazil are in the race of most modern use-case implementation under E-Governance.

1.9   STATUS OF E-GOVERNANCE AROUND THE WORLD

Many of the countries, other than India, have initiated E-Governance programs to enhance the transparency, efficiency and accountability of the government. Some of the recent initiatives taken in other countries have been described as follows.

1.9.1   USA

The 'Expanding Electronic Government' initiative was taken in 2001 aimed at making use of information technology to reduce the governmental paperwork, reduce cost and to speed up the response time to citizens. The federal government intended to modernize the use of IT and use it to focus on different groups such as individuals, businesses, employees and so on. The initial E-Government accomplishments included programs like FirstGov. gov, Volunteer.gov, Recreation.gov, GovBenefits.gov, IRS Free Filing, GoLearn.gov, BusinessLaw.gov, E-Payroll, Regulations.gov and E-Clearance.

The initial efforts regarding E-Governance led to the need of 'Federal Enterprise Architecture' (UNPAN 2003).

1.9.2   UK

The E-Government document 'A Strategic Framework for Public Services in the Information Age' in 2000 marked the emergence of E-Governance in UK. The strategy intends to build services around citizens' choice, make the governmental services available through internet, mobile phones, TV and personal computers. It focuses on social inclusion and using the information better. The framework policies aim to achieve standardization and confidence. The other policies introduced under E-Governance framework were security framework policy, authentication framework policy, smart card framework policy, privacy framework and interoperability framework policy. In UK, E-Envoy, Central IT Unit (CITU) and public sector organizations constitute the institutional mechanism for implementing the strategy (Cabinet Office 2007).

1.9.3   New Zealand

The New Zealand government established an E-Government Unit realizing the opportunities offered by ICT. In April 2001, E-Government strategy was introduced in the country with the vision "By 2004, the

Internet will be the dominant means of enabling ready access to government information, services and processes." The E-Government strategy aimed at better services, cost effectiveness, leadership, improved participation and reputation (Gupta and Jaroliya 2008).

1.9.4   *The UN Survey*

The United Nations, in their United Nations' E-Government Development Index (EGDI) Survey presented a survey report based on the E-Governance preparedness levels internationally. The UN Survey (2008) has assessed the preparedness of E-Governance in various countries based on a comprehensive E-Government readiness index. The index includes components like the web measure index, the telecommunication infrastructure index and the human capital index. These components of the index are discussed below:

1.9.4.1   *The Web Measure Index*
It is based on country's level of online presence and usage in previous years. It is a five-stage model.

1.9.4.2   *The Telecommunication Infrastructure Index*
This index consists of five primary indices that determine the country's infrastructure capacity related to the delivery of E-Government services. These are:

(i) Internet Users/100 persons
(ii) PCs/100 persons
(iii) Main Telephones Lines/100 persons
(iv) Cellular telephones/100 persons
(v) Broad banding/100 persons

1.9.4.3   *The Human Capital Index*
The human capital index is made up of two elements—the adult literacy rate and the combined primary, secondary and tertiary gross enrolment ratio, where two-thirds weight is given to the adult literacy rate and one-third is given to the gross enrolment ratio.

CHAPTER 2

E-Governance in India

Sunil K. Muttoo

2.1 E-Governance Development in India

India is considered as an early adopter of E-Governance among other developing countries. The first wave evolved bottom-up. It was the recommendation of some social entrepreneurs to the district level officials about the wonders of new ICTs that made them consider using IT in government departments. ICT can help providing convergent services to remote areas, and improving transparency (Singh 2012). With the advent of World Wide Web, striking developments were witnessed in IT applications by governments. The technology and E-Governance initiatives have been increasing since then. The increasing recognition of importance of electronics made the government of India to establish a department of Electronics in 1970. Another step taken towards E-Governance was establishment of National Informatics Centre (NIC) in 1977 which brought information and communication in limelight. It was the first major step taken by the government of India that contributed towards E-Governance. However, E-Governance program in India gained prominence with the launch of NICNET in 1987 satellite-based computer network nationwide. Computerization (provision of free hardware and software) was done in all the district, State and National level offices of the country (Mitra 2012).

In India, E-Governance was started in Kerala by the name of *AKSHAYA*. This project provided power of networking and connectivity to 1000–3000 families in Kerala by setting up around 5000 multipurpose community technology centers called *Akshaya E-Kendras* across Kerala. Initially owned

© The Author(s) 2019
S. K. Muttoo et al., *E-Governance in India*,
https://doi.org/10.1007/978-981-13-8852-1_2

by private entrepreneurs, each E-Kendra set up was within 2–3 kilometers of every household. The project Akshaya focused on E-Learning, E-Transaction, E-Governance, information and communication and provides social and economic power to the public (Mitra 2012).

The *Gyandoot* project was launched in 2000 in Dhar district for mass-based information revolution. The purpose of this project was to provide user-charge-based services to the masses and to fulfill the information technology-related developmental needs of government departments and Panchayats without any cost. The aim was to create pressure from the community front-end for digitization of backend departmental processes.

One of the most organized and successful efforts of E-Governance in India made during 2000–2005 was *rural E-Seva* launched in West Godavari district of Andhra Pradesh. The two-community level E-Governance initiatives, *N-logue* and *Drishti* running thousands of community tele-centers across the country to deliver E-Governance services were also successful steps taken by the Indian government.

The early efforts are seldom scaled up. Around 2005–2006, N-logue and *Drishti* were closed. Rural E-Seva also moved out of E-Governance services. However, these initiatives developed a lasting impression of new ICTs as a medium to provide better services to the people as well as make the system more transparent and accountable. They created the framework for the very determined E-Governance project National E-Governance Plan (NeGP), launched by the Government of India in 2006 (Singh 2012).

2.2   STRUCTURE OF E-GOVERNANCE IN INDIA

The institutional mechanism guiding E-Governance in India is headed by an apex body that comprises of senior strategic members from NIC, NASSCOM, IT, BIS, MAIT, CDAC and Planning Commission, the apex body approve, notify and enforce standards. The next level bodies include NIC, DIT that are responsible for standards formulation and STQC which is meant for execution. Expert committees and Working Groups under NIC and Specialist Committees under DIT are formed for planning and formulation of standards and committee formed under STQC release and maintain standards (DeitY 2016a, b, c) (Fig. 2.1).

Institutional Mechanism

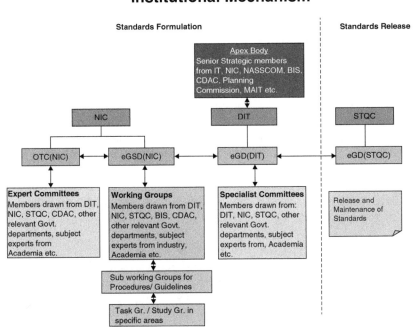

Fig. 2.1   Institutional mechanism of E-Governance. (Source: DeitY 2016a, b, c)

2.3   NATIONAL E-GOVERNANCE PLAN FRAMEWORK

2.3.1   *Initiatives*

2.3.1.1 *National E-Governance Plan*

Numerous initiatives under the E-Governance plan have been taken by the State and Central governments over the years. The initiatives aim at improving the delivery of public services and their assessment process. From computerization of government departments, E-Governance has evolved to advanced services such as citizen centricity, service orientation and transparency. Previous E-Governance initiatives have become a source of learning to shape progressive E-Governance strategy of the country. A programmed approach has been adopted with a vision and strategy to speed up the process of E-Governance at Local, State and

National levels. This approach has been introduced with the aim of reducing the costs, allow interoperability through standards and provide services to the citizens.

The National E-Governance Plan (NeGP) is a comprehensive view of E-Governance initiatives in India. NeGP, initiated by Department of Electronics and Information Technology (DEITY) and Department of Administrative Reforms and Public Grievances (DARPG), aims to provide better government services to the citizens and businesses of the country.

2.3.1.2 Implementation Strategy, Approach and Methodology of NeGP
The implementation process of E-Governance involves provisioning of hardware and software, networking, process re-engineering and change management that makes it a complex process. The earlier experience from E-Governance initiatives, the NeGP has been introduced with the following methodology (India.gov.in 2015).

a. Common Support Infrastructure
   The proper implementation of NeGP requires setting up of support IT infrastructure such as Common Services Centres (CSCs), State Wide Area Networks (SWANs), State Data Centers (SDCs) and Electronic Service Delivery Gateways.
b. Governance
   The process of implementation of the program involves arrangements for monitoring and coordinating the direction of the competent authorities. The proper implementation of the NeGP program requires laying down standards and policy guidelines, providing infrastructure, building necessary capacity, R&D, and so forth. DEITY and other institutions like NIC, STQC, CDAC, and NISG play an effective role in the implementation process and, therefore, are required to strengthen themselves to promote effectiveness.
c. Centralized Initiative, Decentralized Implementation
   A centralized approach is being used to promote E-Governance in order to ensure citizen-centric orientation and optimal utilization of ICT infrastructure and to enhance interoperability of E-Governance applications while promoting a decentralized implementation model. The program intends to identify successful projects and replicating them with modifications as needed by the project.

d. Public-Private Partnerships (PPP)
  The PPP model is suggested to enlarge the resource pool and to deal with security aspects.
e. Integrative Elements
  In order to facilitate integration and avoid ambiguity, unique identification codes have been adopted for citizens and businesses.
f. Program Approach at the National and State Levels
  NeGP needs involvement of various Union Ministries/Departments and State Governments. The program was implemented with well-defined roles and responsibilities of each agency involved for overall integration at the national level. Program management structures have been introduced to implement this.
g. Facilitator Role of DEITY
  DEITY provides support and technical assistance for the implementation of NeGP and is, therefore, considered to be the facilitator of the implementation of the program. It assists the Apex Committee in managing the program. DEITY is considered a catalyst in promoting and implementing pilot/infrastructure/technical/special projects and supporting the program. Government Process Re-engineering and Change Management is another function of DEITY. The NeGP is running from the funds of Planning Commission and Ministry of Finance and through Plan and Non-plan budgetary provisions (DeitY 2016a, b, c).
h. Ownership of Ministries
  Mission Mode Projects have been introduced under NeGP by the concerned line Ministries. E-Governance is being used for major projects like Bharat Nirman, Rural Employment Guarantee Schemes, and so on. The State governments are also given the flexibility to identify a few state-specific projects to contribute to the economic development of the State.

  The NeGP comprises 27 Mission Mode Projects and 8 components as on 18 May 2006. The MMPs are implemented by Central and State Ministries to align with the objectives of NeGP. The infrastructure components include State Data Centers (SDCs), Common Services Centres (CSCs) and National E-Governance Service Delivery Gateway (NSDG), State Wide Area Networks (SWAN), State E-Governance Service Delivery Gateway (SSDG), and Mobile E-Governance Service Delivery Gateway (MSDG) (Kumar 2016).

2.3.2   National E-Governance Division

National E-Governance Division (NeGD) was proposed to enhance proper implementation of NeGP and provide a suitable institutional mechanism to enable DeitY to engage competent resources from the open market and from the Government. NeGD was formed by DeitY in 2009 as an Independent Business Division (IBD) within Media Lab Asia (MLAsia) (a Section 8 Company under DeitY).

The functions of NeGD include Program Management of NeGP, that is, supporting DIT in fulfilling the responsibilities assigned to DIT under NeGP. It helps in proper implementation of NeGP by various Ministries and State Governments and provides technical assistance to Central Ministries and State Line Departments. It serves as a secretariat to the Apex Committee and handles issues related to technology, security, standards and infrastructure. It frames policies, creates organization structure and is responsible for human resource development, training and awareness building.

NeGD is the facilitator for initiatives under Mission Mode Projects and is a supporter for components under NeGP 2.0 across the country. NeGD also supports central and state governments/ministries to implement their E-Governance initiatives (NeGD 2016a, b).

2.3.3   Services

The National E-Governance Plan aims to bring public services closer home to citizens, as stated in its Vision Statement. The 31 Mission Mode Projects (MMPs) launched under the National E-Governance Plan are responsible for Electronic Service Delivery.

Public service is defined as government services provided directly or through any service provider; such as the receipt of forms and applications, issue or grant of any license, permit, or approval and the receipt or payment of money by government.

Electronic Service Delivery is the delivery of government services through electronic mode such as the receipt of forms and applications, issue or grant of any license, permit, or approval and the receipt or payment of money by the government (DeitY 2016a, b, c).

Three kinds of services offered by the Government are as follows:

(a) Government to Citizen (G2C) Services accessed by the Citizens
(b) Government to Business (G2B) Services accessed by the Businesses
(c) Government to Government (G2G) Services accessed by Government Departments

2.3.4   Projects

2.3.4.1 Mission Mode Projects

Under the National E-Governance Plan (NeGP), mission mode projects (MMPs) have been introduced that focuses on one aspect of electronic governance, such as banking, land records or commercial taxes and so on. Mission mode projects are individual projects with clearly defined objectives, scopes, and implementation timelines and milestones and have measurable outcomes and service levels.

Mission mode projects (MMPs) are classified as the state, central or integrated projects specifying their individual needs (DeitY 2016a, b, c) (Fig. 2.2).

The integrated Mission Mode Projects include E-Procurement, E-Courts, E-Biz and common services centres. The recent initiative introduced under Digital India include Direct Cash transfer, MyGov Citizen portal, E-Kranti

Central MMPs	State MMPs	Integrated MMPs
• Banking • Central Excise & Customs • Income Tax (IT) • Insurance • MCA21 • Passport • Immigration, Visa and Foreigners Registration & Tracking • Pension • E-Office • Posts • UID	• Agriculture • Commercial Taxes • E-District • Employment Exchange • Land Records (NLRMP) • Municipalities • E-Panchayats • Police (CCTNS) • Road Transport • Treasuries Computerization • PDS • Education • Health	• CSC • E-Biz • E-Courts • E-Procurement • EDI For ETrade • National E-Governance Service Delivery Gateway • India Portal

Fig. 2.2 Mission Mode Projects. (DeitY 2016a, b, c)

scheme, Digital Cloud, Digi Locker, E-Sign framework, Swachh Bharat Mission mobile App, National Scholarship Portal, E-Hospital, Bharat Net, Wi-Fi hotspots, Next generation network, electronics development fund, Centre of Excellence on Internet of Things (IoT), M-Governance and Mobile Seva (Pareek 2015).

2.3.5   Capacity-Building Scheme

The government of India introduced Capacity-Building Scheme for establishment of institutional framework for strategic decisions at the State Level. The scheme was approved by the Cabinet Committee on Economic Affairs (CCEA) in 2008 with a budget of Rs. 313 Crores. The objectives of the scheme include:

- Setting up of State E-Governance Mission Team (SeMT) for decision making.
- Providing training and set up orientation program to SeMTs and decision makers.
- Setting up of Capacity-Building Management Cell to coordinate and implement plans under the scheme.

Since most states have inadequate personnel and the skill-sets needed to solve the problems, so to improve service orientation in order to fulfill the objectives of NeGP, a capacity-building scheme is necessary to develop skills and handle the challenges. The objectives of capacity-building scheme are to align project design to NeGP service orientation, to enhance consistency across initiatives, allow change management, Government Process Re-engineering, promote resource utilization, leverage external resources, implement and monitor best practices (DeitY 2016a, b, c).

2.3.6   Awareness and Communication

The proper implementation of National E-Governance Plan involves awareness and communication. The awareness and communication efforts of the government help in raising the level of awareness about NeGP, related services and service delivery channels amongst stakeholders across the country.

The Department of Electronics and Information Technology is responsible for the awareness task, for branding of NeGP, its Mission Mode Projects, and improving visibility of E-Services through mass media, rural outreach campaign, conferences, workshops and exhibitions (DeitY 2016a, b, c).

2.3.7   Standards, Policies and Frameworks

Standards ensure sharing of information and interoperability of data across departments and with citizens through E-Governance applications. DeitY is responsible to promote the usage of standards to avoid any technology issues. NeGP has set up an Institutional Mechanism to adopt Standards for E-Governance for areas like Metadata & Data, Localization and Language Technology, information security and interoperability, biometrics, and quality and documentation (DeitY 2016a, b, c) (Fig. 2.3).

2.3.8   Impact and Outcomes

National E-Governance Plan is viewed to promote effective service delivery to citizens and improve the quality of basic governance. A large amount of resources is invested in promoting E-Government projects with the aim to enhance service quality and delivery and reduce challenges related to governance. An assessment strategy is devised for the existing E-Government projects for project appraisal and capacity building. Under the assessment process, DeitY also undertakes E-Readiness Assessment of States and Union Territories (DeitY 2016a, b, c).

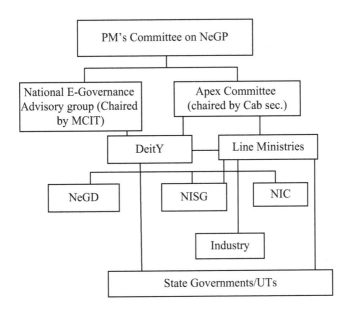

Fig. 2.3 Institutional framework of NeGP at National level. (Source: DeitY 2016a, b, c)

2.4   E-GOVERNANCE INFRASTRUCTURE

The government has realized the importance of IT in growth and therefore has directed all government departments to allocate up to 3 percent of their annual budget to computerization, that is, for procurement of hardware and software and 5 percent for building IT infrastructure (TCS 2008).

2.4.1   NeGP Infrastructure Plan

Under NeGP, core infrastructure and institutional mechanisms have been set up to create an environment favorable to citizens and businesses. The plan involves setting up of State-wide Area Networks (SWAN), State Data Centers (SDCs) and Common Service Centres (CSCs). The National Data Centers (NDCs) were introduced as the core infrastructure under the plan (Fig. 2.4). The four main pillars of NeGP infrastructure initiative are (NeGP 2016) as follows.

1. State-wide Area Networks (SWANs).
2. State Data Centers (SDCs).
3. Common Services Centres (CSCs).
4. Electronic Service Delivery Gateways.

Fig. 2.4 NeGP infrastructure. (NeGP 2016)

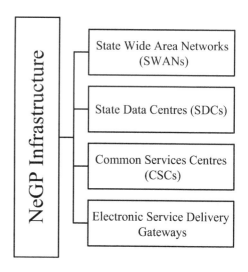

SWANS are the secured networks in the State headquarters up to block level with minimum bandwidth capacity of 2 Mbps. The plan also aimed at establishing 100,000 broadband internet-enabled CSCs in rural areas to provide E-Services to the citizens at their doorsteps.

Many States have established SWANS, data banks and customized applications. The State Data Centers have been established to improve the interconnectivity of the servers. VSAT connectivity has also been introduced in all districts of the country.

An SDC serves as a physical storehouse of public and private data. It acts as a centralized database for various E-Government applications. The purpose of SDC is to provide shared, secured and managed infrastructure for E-Government applications. With the support of State-wide Area Network (SWAN) and Common Service Centre (CSC), SDC is capable of providing efficient electronic delivery of G2G, G2C and G2B services.

CSCs are access points for service delivery that provide high-quality and cost-effective data content and services to support E-Governance. It provides services in the areas of E-Governance such as agriculture, education, banking, health, commercial, telemedicine, entertainment as well as other private services. It promotes rural entrepreneurship, builds capacities, enables community participation and brings social change (NeGP 2016).

Electronic Service Delivery Gateways are messaging and routing switches that allow interoperability and exchange of electronic data across government departments. It helps in disseminating government information anytime and maintains online records of the same. It enables status tracking, downloading of forms and application submission by the citizens. Mobile service delivery gateway (Mobile Seva) is another platform for delivery of services. It makes use of mobile phones that provide multiple mobile-based channels SMS, USSD, IVRS, m-Apps to reach citizens of the country especially in rural areas.

The purpose of MSDG is to deliver Government services over mobile devices using mobile applications that are installed on the user's mobile handsets. The users generally include backend departments and citizens. It allows integration with backend department with the help of NSDG/SSDG, E-Governance exchange infrastructure.

MSDG offers services like SMS Gateway, Interactive Voice Response Systems (IVRS) based Services, and Unstructured Supplementary Services Data (USSD) based services and Mobile Applications and M-Gov Application Store (NeGP 2016).

2.4.2   Other Initiatives under NeGP

Apart from the main four pillars of NeGP infrastructure initiative, many other initiatives have been taken under the scheme to improve the infrastructure (Fig. 2.5).

Another component of the plan included Capacity-Building Scheme to establish a mechanism for capacity building and training of the end user. Capacity-Building Management Cell at the center and State E-Governance Mission Team have been set up for this purpose.

Another important infrastructure initiative of the plan includes single-window access to all government-related information and services at all levels from Central government to State government to district administrations and Panchayats.

Satellite-based country-wide communication network (NICNET) has been initiated to provide connectivity in various Central, State and District ministries/departments which provides linkages in 611 districts with 3000 nodes in Wide Area Network (WAN) and Local Area Network (LAN) (Poulose 2010).

A very important initiative in this regard is E-Taal (Electronic Transaction Aggregation and Analysis Layer). E-Taal is a web portal developed for dissemination of statistics related to E-Transactions of E-Governance Projects including Mission Mode Projects. It collects transaction statistics from web-based applications and presents transaction counts done by various E-Governance projects in tabular and graphical form (Swathi 2016).

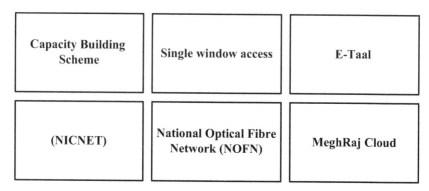

Fig. 2.5   Other initiatives under NeGP Infrastructure Scheme. (NeGP 2016)

National Information Infrastructure (NII) is another initiative of the government to improve the infrastructure and is responsible for integration of network and cloud infrastructure with the objective of providing speedy connection and cloud platform to government departments at all levels. The infrastructure components of the initiative are State-wide Area Network (SWAN), National Knowledge Network (NKN), National Optical Fibre Network (NOFN), Government User Network (GUN) and the MeghRaj Cloud (Swathi 2016).

2.4.3   E-Governance under Digital India

The infrastructure plans under Digital India is another initiative to support E-Governance in India. One of the aims of Digital India plan is to provide Infrastructure as Utility to every citizen. It involves the private sector with the public sector to provide last mile access, location-specific Wi-Fi access (schools, universities, etc.) and applications to provide cloud-based services on demand to citizens such as branchless banking, skill development, health and education and E-Justice. Social Media, Mobility, Analytics and Cloud (SMAC) are visions of digital India campaign to provide governance and services on demand and digitally empower citizens.

E-Governance Policy Initiatives under Digital India include the E-Kranti Framework, Open Source Software, Open APIs, E-Mail Policy, Use of IT Resources, Collaborative Application Development and Application Development and Re-Engineering for Cloud Ready Applications. This compendium of initiatives intends to provide help to policy makers and practitioners to fasten the implementation of Digital India projects (Digital India 2016a, b, c).

CHAPTER 3

UN E-Government Report: Status of India

Rajan Gupta

3.1 UN E-Government Report

E-Governance has evolved over the years from computerization to initiatives that reflect finer points of governance such as transparency and service orientation. However, it is important to know whether it is performing at the pace other countries are developing. A comparison needs to be made among different countries based on their governance to assess the current performance and to know what improvements are required in the E-Governance process. This comparison has been made by the United Nations through their E-Governance survey.

The United Nations E-Government Survey presents a comparison of 193 United Nations Member States based on E-Government development status. It highlights the progress of E-Government in countries measured on three important dimensions (i) *availability of online services*, (ii) *telecommunication infrastructure* and (iii) *human capacity*. The survey helps the decision makers to identify the strengths and challenges in E-Government and guides the development of E-Government policies and strategies. The survey focuses on emerging E-Governance trends, issues, opportunities and innovative practices of E-Government development. The beneficiaries of the report include government officials, intergovernmental institutions, academics, private sector and citizens at large. This chapter aims to compare the E-Governance development in India over the years on various aspects and with other countries' E-Governance development. It describes the present status and improvement of E-Governance in India (Table 3.1).

© The Author(s) 2019
S. K. Muttoo et al., *E-Governance in India*,
https://doi.org/10.1007/978-981-13-8852-1_3

Table 3.1 E-Government development index (2014)

Rank	Country	EGDI
1	Republic of Korea	0.9462
2	Australia	0.9103
3	Singapore	0.9076
4	France	0.8938
5	Netherlands	0.8897
6	Japan	0.8874
7	United States of America	0.8748
8	United Kingdom of Great Britain and Northern Ireland	0.8695
9	New Zealand	0.8644
10	Finland	0.8449
117	Namibia	0.3880
118	**India**	**0.3834**
119	Kenya	0.3805
185	South Sudan	0.1418
186	Sierra Leone	0.1329
187	Central African Republic	0.1257
188	Papua New Guinea	0.1203
189	Chad	0.1076
190	Guinea	0.0954
191	Niger	0.0946
192	Eritrea	0.0908
193	Somalia	0.0139

Source: UN E-Government Report (2014a, b)

A cross-country comparison has been done based on their E-Governance development level. The UN report on E-Governance (2014a, b) has ranked India 118 in the EGDI out of 193 ranks which shows that India has a low E-Governance level as in lesser online presence, weak telecommunication infrastructure and inefficient human capital as compared to other countries. India's EGDI value as indicated by the UN report 2014 is 0.3834. The top ten countries with highest EGDI are Republic of Korea, Australia, Singapore, France, Netherlands, Japan, USA, UK, New Zealand and Finland. Countries having EGDI similar to India includes Namibia, Cuba, Belize and Kenya. There are many countries having EGDI lower than that of India. Some of these are Guinea, Niger, Eritrea and Somalia (Table 3.2).

Table 3.2 depicts the top ten countries at the highest position in the EGDI list over the years (2005-2014). It shows the change in ranks of countries based on EGDI over a decade. Republic of Korea has been at the top of the list since 2010 replacing Sweden and United States of America.

Table 3.2 Top ten performers based on EGDI (2005–2014)

Rank	2014	2012	2010	2008	2005
1	Republic of Korea	Republic of Korea	Republic of Korea	Sweden	United States of America
2	Australia	Netherlands	United States of America	Denmark	Denmark
3	Singapore	United Kingdom	Canada	Norway	Sweden
4	France	Denmark	United Kingdom	United States	United Kingdom
5	Netherlands	United States of America	Netherlands	Netherlands	Republic of Korea
6	Japan	France	Norway	Republic of Korea	Australia
7	United States of America	Sweden	Denmark	Canada	Singapore
8	United Kingdom of Great Britain and Northern Ireland	Norway	Australia	Australia	Canada
9	New Zealand	Finland	Spain	France	Finland
10	Finland	Singapore	France	United Kingdom	Norway
	India (118)	India (125)	India (119)	India (113)	India (87)

Source: UN E-Government Report (2014a, b)

Australia, Singapore, Japan and New Zealand have been observed to undergo significant improvement in governance. According to the UN survey 2014, there has been a significant improvement in the ranks of Australia, Singapore, Japan and New Zealand. The rank in case of India, however, has been declined over the years.

3.2 Components of E-Governance

The E-Governance status of the countries is determined through their E-Participation, online service, telecommunication infrastructure and human capital index (Table 3.3).

The E-Government development index of India over the years has been presented (2005–2014). The EGDI values have been quite fluctuating decreased in initial years (2005–2010) and then improved (2010–2014). The rank is been accordingly declined from the year 2005 to 2012. It is now lowered in the year 2014 as compared to the year 2012 (Table 3.4).

Table 3.3 E-Government
development index (India)
(2005–2014)

Year	Rank	EGDI
2014	118	0.3834
2012	125	0.3829
2010	119	0.3567
2008	113	0.3814
2005	87	0.4001

Table 3.4 E-Participation index (Top 50 performers in 2014)

1. Netherlands	11. Colombia	21. Spain	31. Norway	41. Belgium
2. Republic of Korea	12. Israel	22. Estonia	32. Russian Federation	42. India
3. Uruguay	13. United Arab Emirates	23. Kazakhstan	33. China	43. Republic of Moldova
4. France	14. Bahrain	24. Brazil	34. Ireland	44. Slovakia
5. Japan	15. Canada	25. Finland	35. Kenya	45. El Salvador
6. United Kingdom	16. Costa Rica	26. Germany	36. Lithuania	46. Mexico
7. Australia	17. Greece	27. Latvia	37. Portugal	47. Qatar
8. Chile	18. Morocco	28. Oman	38. Sri Lanka	48. Sweden
9. United States of America	19. Italy	29. Peru	39. Tunisia	49. Georgia
10. Singapore	20. New Zealand	30. Mongolia	40. Austria	50. Montenegro

Source: UN E-Government Report (2014a, b)

3.3   E-PARTICIPATION IN THE WORLD

The UN report shows the E-Participation level of the countries and the top performers have been depicted in Table 3.5, India being one of them. The top ten nations having highest E-Participation level include Netherlands, Republic of Korea, Uruguay, France, Japan, United Kingdom, Australia, Chile, USA and Singapore. India stands at the 42nd position at the E-Participation index. Among the top 50 countries, 28 are high income, 14 belong to upper middle income and 7 countries Morocco, Mongolia, Sri Lanka, India, Republic of Moldova, El Salvador and Georgia are lower middle income. Only one of these countries (Kenya) is a low-income country.

Table 3.5 E-Participation
in India (2005–2014)

Year	Rank	Index value
2005	34	0.1587
2008	49	0.25
2010	58	0.2
2012	25	0.1842
2014	40	0.6275

(UN E-Government Report 2005, 2008, 2010, 2012, 2014a, b)

3.4   E-Participation in India

The E-Participation level in India over the years has also been assessed. The E-Participation level has risen significantly in the year 2014 as compared to the previous years. The reason might be the introduction of various E-Governance initiatives and Digital India campaign launched by the new government elected in the corresponding years. A drastic change has been witnessed in the participation in year 2014 (0.6275) as compared to 2012 (0.1842) although the rank has declined from 25 to 40.

Table 3.6 shows India's level of E-Governance through analyzing its components—human capital, telecommunication infrastructure and online service. The online service component has declined from the level of year 2005 and then improved since 2010. The online service component index in the year 2014 is 0.5433. The telecommunication infrastructure component has increased in 2008 from 2005 but has declined since then. The human capital index has been fluctuating over the years. The lowest human capital level was witnessed in the year 2010 while the highest in the past decade was found to be in 2008 (Table 3.7).

Table 3.8 compares the telecommunication infrastructure of India of the year 2012 and 2014. The telecommunication infrastructure has improved in 2014 (0.1372) as compared to 2012 (0.1102) as per the UN E-Governance survey. The estimated internet users have increased from 7.5 percent in 2012 to 12.58 percent in 2014. Main fixed phone lines have reduced from 2.87 percent to 2.51 percent in two years. Mobile subscribers have increased from 61.42 percent in 2012 to 69.92 percent in 2014. The fixed internet subscriptions have reduced from 1.53 percent to 1.16 percent in two years while users of fixed broadband have increased drastically from 0.90 percent in 2012 to 4.99 percent in 2014 (United Nations 2012, 2014a, b).

Table 3.6 Components of E-Governance (India)

Year	Online service component	Telecommunication infrastructure component	Human capital component
2014	0.5433	0.1372	0.4698
2012	0.5359	0.1102	0.5025
2010	0.1252	0.0192	0.2123
2008	0.4783	0.0435	0.6195
2005	0.5827	0.0277	0.5900

Source: UN E-Government Report (2014a, b)

Table 3.7 Online service index and its components (India)

Year	Online service Index value	Stage I %	Stage II %	Stage III %	Stage IV %	Total %
2012	0.5359	100	64	33	38	47
2014	0.5433	97	59	21	29	50

Source: UN E-Government Report (2014a, b)

Table 3.8 Telecommunication infrastructure index and its components (India)

Year	Index value	Estimated Internet users	Main fixed phone lines	Mobile subscribers	Fixed Internet subscriptions	Fixed broadband
2012	0.1102	7.50	2.87	61.42	1.53	0.90
2014	0.1372	12.58	2.51	69.92	1.16	4.99

Source: UN E-Government Report (2014a, b)

The human capital index and its components (adult literacy and enrollment) in India have been assessed and the human capital index is found to have been reduced from 0.5025 in 2012 to 0.4698 in 2014. The adult literacy component has been stable in the two years (62.75 percent) while enrollment ratio increased from 62.61 percent in 2012 to 65.07 percent in 2014 (United Nations 2012; 2014a, b) (Table 3.9).

India is among the top 50 countries with a score higher than 66.6 percent in data publishing (Table 3.10).

The ranks based on E-Governance Development Index (EGDI) (United Nations 2014a, b) of BRICS nations have been presented. The data suggests a vast difference in the E-Governance level of BRICS nations. Russia leads in E-Governance level among the BRICS nations followed by Brazil, China and South Africa while India ranks the lowest (Table 3.11).

Table 3.9   Human capital index and its components (India)

Year	Human capital index	Adult literacy (%)	Enrollment (%)
2012	0.5025	62.75	62.61
2014	0.4698	62.75	65.07

Source: UN E-Government Report (2014a, b)

Table 3.10   Countries with a score higher than 66.6 percent in data publishing

Albania	Denmark	Italy	Netherlands	Singapore
Australia	El Salvador	Japan	New Zealand	Spain
Austria	Estonia	Kazakhstan	Norway	Sri Lanka
Bahrain	Finland	Kenya	Oman	Sweden
Belgium	France	Latvia	Peru	Thailand
Brazil	Georgia	Lithuania	Portugal	Tunisia
Canada	Germany	Luxembourg	Qatar	United Arab Emirates
Chile	India	Malta	Republic of Korea	United Kingdom
China	Ireland	Mexico	Republic of Moldova	United States of America
Costa Rica	Israel	Morocco	Saudi Arabia	Uruguay

Source: UN E-Government Report (2014a, b)

Table 3.11   E-Governance in BRICS Nations

Country	Position
Brazil	57
Russia	27
India	118
China	70
South Africa	93

Source: UN E-Government Report (2014a, b)

3.5   WASEDA-IAC INTERNATIONAL E-GOVERNMENT SURVEY

Apart from UN E-Government survey, WASEDA-IAC International E-Government surveys also provide rankings to the countries based on their E-Governance performance (Table 3.12).

Table 3.12   Waseda-IAC International E-Government ranking 2015

No.	Total rankings	Score	No.	Total rankings	Score	No.	Total rankings	Score
1	Singapore	93.80	22	Thailand	67.31	43	Brunei	51.06
2	USA	93.58	23	Israel	65.80	44	Bahrain	50.50
3	Denmark	91.25	24	HK SAR	65.24	45	Brazil	50.37
4	UK	90.17	25	Malaysia	64.87	46	Argentina	50.32
5	Korea	89.39	26	Portugal	63.93	47	Colombia	49.36
6	Japan	87.77	27	Czech Republic	63.48	48	South Africa	49.30
7	Australia	86.30	28	Italy	61.30	49	China	48.36
8	Estonia	84.87	29	Indonesia	60.11	50	Kazakhstan	47.73
9	Canada	81.45	30	UAE	58.10	51	Saudi Arabia	47.48
10	Norway	79.63	31	Poland	57.30	52	Peru	46.21
11	Sweden	77.95	32	Spain	57.12	53	Tunisia	45.87
12	Austria	77.26	33	Vietnam	57.03	54	Venezuela	44.65
13	New Zealand	76.66	34	Russia	56.56	55	Uruguay	44.01
14	Finland	76.49	35	India	56.42	56	Morocco	43.13
15	Germany	76.46	36	Macau SAR	56.27	57	Pakistan	42.94
16	France	73.39	37	Chile	53.49	58	Costa Rica	42.06
17	Chinese Taipei	72.76	38	Mexico	53.41	59	Georgia	40.73
18	Belgium	71.69	39	Romania	53.11	60	Nigeria	38.37
19	Iceland	69.73	40	Oman	51.60	61	Fiji	37.54
20	Netherlands	69.53	41	Philippines	51.47	62	Egypt	37.19
21	Switzerland	69.17	42	Turkey	51.31	63	Kenya	32.91

Source: Toshio (2015)

The Institute of E-Government at Waseda University (Director: Prof. Toshio Obi), Tokyo, in cooperation with the International Academy of CIO (IAC) releases E-Government ranking report since 2004. The 2015 survey presented E-Government development worldwide and given ranks to 63 countries based on their E-Governance performance. The survey suggested Singapore to be the best performer in E-Governance services followed by United States, Denmark, United Kingdom and Korea. India ranks 35 among the 63 countries in E-Governance development (Toshio 2015) (Table 3.13).

Waseda-IAC E-Government Ranking Survey 2015 has assessed the E-Governance level in big population countries. Countries with population higher than 100 million were selected to disclose the difficulty they face in implementing E-Governance policies. Countries with large population have a large territory as well and face developmental challenges in implementing E-Government such as delivering E-Services, building infrastructure to support the E-Governance policies and so on. USA tops

Table 3.13 Ranking in big population countries (higher than 100 million)

No.	Country	Score	No.	Country	Score	No.	Country	Score
1.	USA	93.58	5.	India	56.42	9.	China	48.36
2.	Japan	87.77	6.	Mexico	53.41	10.	Pakistan	42.94
3.	Indonesia	60.11	7.	Philippines	51.47	11.	Nigeria	38.37
4.	Russia	56.56	8.	Brazil	50.37			

Source: Toshio (2015)

Table 3.14 Ranking in Asia-Pacific countries

No.	Country	Score	No.	Country	Score	No.	Country	Score
1.	Singapore	93.80	7.	Thailand	67.31	13.	Macau	56.27
2.	Korea	89.39	8.	HK SAR	65.24	14.	Philippines	51.47
3.	Japan	87.77	9.	Malaysia	64.87	15.	Brunei	51.06
4.	Australia	86.30	10.	Indonesia	60.11	16.	China	48.36
5.	New Zealand	76.66	11.	Vietnam	57.03	17.	Pakistan	42.94
6.	Chinese Taipei	72.76	12.	India	56.42	18.	Fiji	37.54

Source: Toshio (2015)

the list as it keeps up development in E-Government and the government is really committed in delivering public services. India ranks 5th in this list with a score of 56.42. The population is quite high and the country faces challenges such as a lack of transparency and inefficiency. But the recent initiatives have paced the E-Governance development. China, Pakistan and Nigeria are ranked at the bottom of this list (Table 3.14).

India ranks 12th with a score of 56.42 among the 52 countries and territories in the Asia-Pacific region in terms of E-Governance. The rank of India has dropped from last year. In the Asia-Pacific region, Singapore tops the list of E-Governance and Pakistan and Fiji lie in the bottom of this group (Toshio 2015).

CHAPTER 4

Regional Analysis of E-Governance in India

Rajan Gupta

The National Council of Applied Economic Research (NCAER) is responsible for preparing E-Readiness report for Indian states for the last 15 years (since 2003) for ranking Indian states and Union Territories on the basis of their electronic readiness. The conceptual structure with different measures has been kept same over the years. However, research methodology for E-Readiness assessment having different variables have been changed for betterment over the last 15 years. The ranking given to the states based on the E-Readiness are not strictly comparable with previous years, since the measures used in creating the E-Readiness index of different years were different.

4.1 The E-Readiness Index

E-Readiness index (2008) was developed with the help of components like Usage, Readiness and Environment. Usage and Environment were given similar weights while readiness was given a slighter weight (Table 4.1). The states have been classified in the following categories based on the index values of E-Readiness.

(a) Leaders (L1)
(b) Aspiring leaders (L2)
(c) Expectants (L3)
(d) Average achievers (L4)
(e) Below-average achievers (L5)
(f) Least achievers (L6)

© The Author(s) 2019
S. K. Muttoo et al., *E-Governance in India*,
https://doi.org/10.1007/978-981-13-8852-1_4

Table 4.1   Ranking of states

Ranking	States
Leaders	Tamil Nadu, Maharashtra, Karnataka, Andhra Pradesh, Chandigarh, Delhi
Aspiring Leaders	Gujarat, West Bengal, Punjab, Haryana, Kerala
Expectants	Assam, Uttar Pradesh, Goa, Himachal Pradesh, Madhya Pradesh, Andaman and Nicobar, Orissa, Bihar
Average Achievers	Jharkhand, Chhattisgarh, Rajasthan, Uttarakhand, Sikkim
Below-Average Achievers	Tripura, Meghalaya, Puducherry, Nagaland
Least Achievers	Manipur, Jammu and Kashmir, Mizoram

Source: DIT & NCAER (2010)

The leaders and aspiring leader's category include three major blocks: one is south with four states, three out of which are leaders and one, that is, Kerala, is an inspiring leader. The second block lies in the western region with Maharashtra as a leader and Gujarat being the aspiring leader. Third block corresponds to the north-west region where Chandigarh and Delhi are parts of leader group, while aspiring leaders of the region are Punjab and Haryana. The only aspiring leader from the eastern region is West Bengal. The category of under achievers and least achievers majorly involves north-eastern states and some Union Territories. The new states Chhattisgarh, Uttarakhand and Jharkhand belong to the category of average achievers. These states have advantages in terms of latest technology, but they would take time to develop as the regions under them were not fully developed before they got separated from their respective mother states. This can be seen as the mother states Bihar, Uttar Pradesh and Madhya Pradesh are just a single ranking above them.

The performance of Andaman and Nicobar has been proved commendable finding it in the category of expectants on the E-Readiness Index. This could be due to geographical disadvantage as it has been separated by huge water bodies from the peninsular region of India. And also it does not have any benefit from the neighbor countries still the performance is considered surprisingly well.

The states and UTs have also been classified on the basis of ICT usage, environment, readiness and E-Readiness and on the level of E-Readiness (Table 4.2).

Under Level 1, only Karnataka and Chandigarh, among the leader category, have been found in all three sub-components. Maharashtra ranks high in environment and readiness but performs low in ICT usage which implies that it should use enabling environment and high readiness to improve usage of ICT in the state.

Table 4.2 Ranking of states E-Readiness index (2008)

Levels	Environment	Readiness	Usage	E-Readiness
L1	Chandigarh, Karnataka, Andhra Pradesh, Maharashtra	Chandigarh, Tamil Nadu, Maharashtra, Karnataka	Chandigarh, Andaman and Nicobar, Delhi, Karnataka,	Chandigarh, Karnataka, Delhi, Tamil Nadu, Andhra Pradesh, Maharashtra
L2	West Bengal, Kerala, Tamil Nadu, Assam, Haryana, Delhi, Madhya Pradesh, Punjab, Gujarat, Uttar Pradesh	Delhi, Kerala, Punjab, Haryana, Bihar, West Bengal, Andhra Pradesh	Bihar, Delhi, Punjab, Andhra Pradesh, Haryana, Gujarat, Kerala, Tamil Nadu, West Bengal, Uttarakhand	Kerala, West Bengal, Haryana, Punjab, Gujarat
L3	Goa, Orissa, Tripura, Nagaland, Puducherry	Gujarat, Himachal Pradesh, Andaman and Nicobar, Goa, Madhya Pradesh, Orissa	Madhya Pradesh, Himachal Pradesh, Haryana, Punjab, Sikkim, Assam, Chhattisgarh, Maharashtra, Rajasthan	Andaman and Nicobar, Goa, Uttar Pradesh, Madhya Pradesh, Orissa, Himachal Pradesh, Assam, Bihar
L4	Chhattisgarh, Sikkim, Himachal Pradesh, Rajasthan	Uttarakhand, Sikkim, Jharkhand, Rajasthan, Chhattisgarh, Uttar Pradesh	Goa, Bihar, Uttar Pradesh, Orissa, Meghalaya	Sikkim, Uttarakhand, Chhattisgarh, Jharkhand, Rajasthan
L5	Jharkhand, Manipur, Uttarakhand, Andaman and Nicobar, Bihar, Jammu and Kashmir	Jammu and Kashmir, Puducherry, Meghalaya, Assam, Tripura	Mizoram, Nagaland	Puducherry, Nagaland, Tripura, Meghalaya
L6	Mizoram, Meghalaya, Dadra and Nagar Haveli, Arunachal Pradesh, Lakshadweep, Daman and Diu, Arunachal Pradesh	Mizoram, Meghalaya, Dadra and Nagar Haveli, Arunachal Pradesh, Lakshadweep, Daman and Diu, Arunachal Pradesh	Puducherry, Tripura, Dadra and Nagar Haveli, Manipur, Daman and Diu, Arunachal Pradesh, Lakshadweep, Jammu and Kashmir	Mizoram, Manipur, Dadra and Nagar Haveli, Jammu and Kashmir, Lakshadweep, Daman and Diu, Arunachal Pradesh

Source: DIT & NCAER (2010)

Some of the north-eastern states, such as Tripura and Nagaland, despite having enabling environment, have performed poorly in readiness and usage. This implies that policy implementation should be improved in these regions.

Some states/UTs such as Uttarakhand, Andaman and Nicobar, Jharkhand and Chhattisgarh, on the other hand, have performed extremely well in terms of usage but lacks on the environment and readiness component. This implies with better infrastructure and capacity building, the states may improve on readiness and environment front (Table 4.3).

Table 4.3 Ranking of states year-wise

State and UTs	2003	2004	2005	2006	2008
Karnataka	1	1	3	1	1
Andhra Pradesh	4	3	1	2	6
Chandigarh	8	5	5	3	2
Haryana	15	11	9	4	9
Delhi	7	9	8	5	5
Maharashtra	2	4	6	6	3
Tamil Nadu	3	2	2	7	4
Uttar Pradesh	10	15	12	8	18
Punjab	13	10	7	9	11
Kerala	11	6	4	10	8
Rajasthan	16	20	14	11	14
Gujarat	5	7	11	12	10
West Bengal	9	12	15	13	7
Goa	6	8	10	14	14
Chhattisgarh	19	16	16	15	20
Himachal Pradesh	17	19	17	16	17
Madhya Pradesh	12	14	21	17	13
Jharkhand	26	26	22	18	22
Orissa	20	17	20	19	15
Mizoram	21	21	23	20	30
Puducherry	14	13	13	21	28
Sikkim	30	18	19	22	23
Uttarakhand	18	25	18	23	21
Meghalaya	23	24	24	24	28
Assam	25	23	25	25	16
Nagaland	32	35	32	26	26
Bihar	28	32	28	27	27
Andaman and Nicobar	24	31	31	28	20
Lakshadweep	27	27	26	29	33
Jammu and Kashmir	29	22	27	30	31
Tripura	22	29	33	31	25
Manipur	34	28	29	32	29
Daman and Diu	33	33	35	33	35
Arunachal Pradesh	31	30	30	34	31
Dadra and Nagar Haveli	35	34	34	35	34

Source: DIT & NCAER (2010)

The States/UTs which have experienced a significant improvement in their E-Readiness status from 2005 to 2008 are Chandigarh, Haryana, Jharkhand, Orissa, Sikkim, Assam, Nagaland and Manipur. There are other States/UTs that have shown some improvement in E-Readiness like Dadra and Nagar Haveli, Andaman and Nicobar, Bihar, West Bengal, Rajasthan, Kerala, Punjab, Tamil Nadu, and Delhi. However, some States/ UTs have experienced a decline in the rank based on E-Readiness such as Daman and Diu, Tripura, Jammu and Kashmir, Lakshadweep, Meghalaya, Uttarakhand, Puducherry, Mizoram, Chhattisgarh, Goa, Madhya Pradesh, Gujarat, Uttar Pradesh, Tamil Nadu, Maharashtra and Andhra Pradesh. While there are some States/UTs that have similar ranking in 2008 as it was in 2005 such as Arunachal Pradesh, Himachal Pradesh and Karnataka.

4.2   REGIONAL COMPARISON OF E-READINESS

The regional comparison is done to see the spatial interaction effect of ICT (Fig. 4.1).

Almost all southern states are leaders except the islands with underdeveloped ICT such as in Puducherry and Lakshadweep. Lakshadweep cannot maximize interaction effect properly as it is an island; however, the region of Puducherry does not have such issue and, therefore, should have been in better position. The eastern region states like Orissa, West Bengal and Andaman and Nicobar have performed consistently as most of them are categorized as either 'aspiring leaders' or 'expectants'. Rajasthan is the only region with low performance in the north-west region. Among north-eastern states, Assam has performed better than the other states of its region. The other states are majorly categorized into bottom most group except Sikkim. The western region has performed incongruently, with Dadra and Nagar Haveli and Daman and Diu performing poorly and requiring a lot of attention (Fig. 4.2).

The index value for a particular region is calculated by weighing each state by the ratio of its population to the total population of the region, in a manner that the states with smaller region get lower weights and the states with larger area get higher weights. This is done due to the fact that index value of a state cannot be calculated or computed simply by taking the average of the values for the regions as the different regions are of unequal size. Thus, states like Daman-Diu and Dadra Nagar Haveli are assigned a lower weight in contrast to the state of Maharashtra which has 65 percent population of the whole western region. The computed values for various regions are comparatively less and compressed than the originally computed values

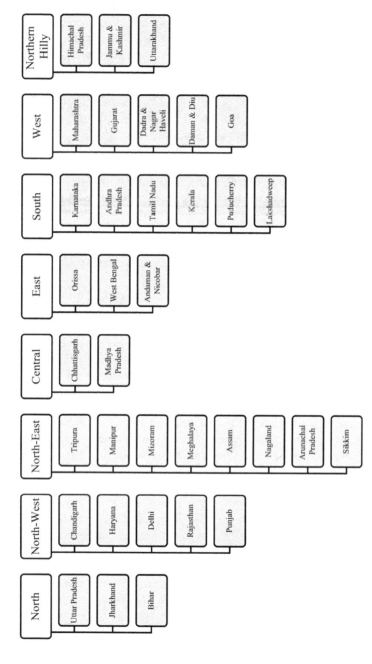

Fig. 4.1 Regional division of states. (DIT & NCAER, 2010)

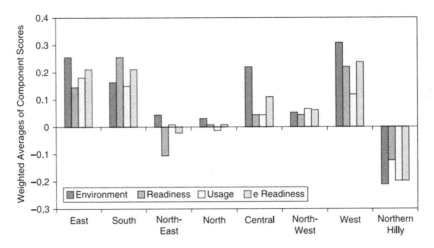

Fig. 4.2 E-Readiness and its components: Regional patterns. (DIT & NCAER, 2010)

for the index. The new values, computed after taking the overall population into consideration, give edge to western region over southern and eastern regions. Favorable environment with respect to the usage and readiness have been observed in the western region after obtaining the weighted average scores. The hilly states like Himachal Pradesh, Jammu and Kashmir and Uttarakhand in the northern region are not doing very well as per the calculations. The states in the north-east regions have low E-Readiness scores and are placed at the second last position with only Assam performing a little better than most of the other states.

4.3 Comparison of E-Readiness Level Over Time

There has been movement of States/UTs across categories for following possible reasons:

- Relative growth in ICT as compared to other states/UTs
- Amendments over the time period in the data variables
- Changes in the state's data for comparative analysis

The changes in the ranking for E-Readiness of the different regions in India and their respective states have been reported for over three years (Table 4.4). The states within different regions which have shown significant

Table 4.4 E-Readiness index for three years (2005, 2006 & 2008)

Levels	E-Readiness			Downward movement (from 2005 to 2008)	Upward movement (from 2005 to 2008)
	4.4. 2005	4.5. 2006	2008		
L1	Chandigarh, Maharashtra, Tamil Nadu, Punjab, Goa	Chandigarh, Kerala, Delhi, Andhra Pradesh, Karnataka, Haryana, Tamil Nadu, Punjab	Tamil Nadu, Karnataka, Maharashtra, Delhi, Chandigarh, Andhra Pradesh	Punjab, Goa	
L2	Haryana, Delhi, Kerala, Andhra Pradesh, Karnataka, Gujarat, Puducherry, Uttar Pradesh	Maharashtra, Gujarat, Goa, Uttar Pradesh	West Bengal, Gujarat, Kerala, Punjab, Haryana	Puducherry, Uttar Pradesh	Karnataka, Andhra Pradesh, Delhi
L3	Sikkim, Chhattisgarh, Himachal Pradesh, West Bengal, Rajasthan, Uttaranchal	West Bengal Himachal Pradesh Chhattisgarh Jharkhand Rajasthan	Madhya Pradesh, Himachal Pradesh, Goa, Assam, Uttar Pradesh, Andaman and Nicobar, Orissa, Bihar	Sikkim, Rajasthan, Chhattisgarh, Himachal Pradesh, Uttaranchal	West Bengal
L4	Meghalaya, Orissa, Jharkhand, Mizoram, Madhya Pradesh, Assam	Madhya Pradesh, Meghalaya, Orissa, Sikkim, Uttarakhand, Puducherry, Mizoram	Chhattisgarh, Uttarakhand, Jharkhand, Sikkim, Rajasthan	Meghalaya, Mizoram	Orissa, Madhya Pradesh, Assam
L5	Bihar, Nagaland, Jammu and Kashmir, Lakshadweep	Andaman and Nicobar Assam Nagaland Lakshadweep Bihar	Tripura, Puducherry, Nagaland, Meghalaya	Jammu and Kashmir, Lakshadweep	Bihar
L6	Tripura, Andaman and Nicobar, Dadra and Nagar Haveli, Arunachal Pradesh, Manipur, Daman and Diu	Arunachal Pradesh, Tripura, Manipur, Dadra and Nagar Haveli, Jammu and Kashmir, Daman and Diu	Mizoram, Daman and Diu, Manipur, Arunachal Pradesh, Lakshadweep, Jammu and Kashmir, Dadra and Nagar Haveli		Andaman and Nicobar, Tripura

Source: DIT & NCAER (2010)

improvement are Bihar and Andaman islands. Following them, states which have shown lesser growth are Andhra Pradesh, Karnataka, Madhya Pradesh, Orissa, Tripura and Assam. The state showing negative growth is Mizoram which have moved down by two levels. Other states showing negative growth include Goa, Punjab, Uttar Pradesh, Puducherry, Chhattisgarh, Sikkim, Meghalaya, Rajasthan, Lakshadweep, and Jammu and Kashmir. Overall, eastern region has shown positive growth while north-east and north-west regions have shown negative growth.

4.4   Components of E-Readiness

The level of ICT environment in States/UTs has been shown in Table 4.5. The movement of the States/UTs across categories of the environment index has been presented for three years. The States and UTs that show relatively upward movement are Tripura, Madhya Pradesh and Bihar as they move up two levels. The states like Assam, West Bengal, Nagaland, Orissa, Andaman and Nicobar, Jammu and Kashmir, Manipur, Andhra Pradesh and Karnataka, have shown very less positive growth. On the other hand, Mizoram, Goa and Uttarakhand have shown a significant downward movement and moved from the level of average achievers (L4) to least achievers (L6). Other states including Punjab, Kerala, Tamil Nadu, Puducherry, Himachal Pradesh, Sikkim, Chhattisgarh, Rajasthan, Meghalaya, and Lakshadweep, have shown a negative growth.

Eastern region has found to be having the most positive growth while north-east region has witnessed a negative growth.

The level of readiness in States/UTs has been shown in Table 4.6. The movement of the various regions for readiness index has been presented for three years. Regions like Andaman and Nicobar and Bihar have shown positive growth as they have moved up by two levels. Regions like Tripura, Madhya Pradesh, West Bengal, Chandigarh, Orissa, Haryana and Karnataka, have shown less positive growth. On the other hand, significant negative growth has been shown by states like Mizoram, Uttar Pradesh and Goa. Other states have also shown some downward mobility including Chhattisgarh, Rajasthan, Gujarat, Andhra Pradesh, Sikkim and Uttaranchal, Assam, Meghalaya, Nagaland and Lakshadweep.

Eastern region has shown positive growth while north-east region has witnessed a downward movement/negative growth.

Table 4.5 Environment index for three years (2005, 2006 & 2008)

Levels	2005	2006	2008	Downward movement (from 2005 to 2008)	Upward movement (from 2005 to 2008)
L1	Goa Chandigarh Punjab Maharashtra Kerala Tamil Nadu	Chandigarh Haryana Maharashtra Gujarat Delhi Punjab	Karnataka Chandigarh Andhra Pradesh Maharashtra	Goa Punjab Kerala Tamil Nadu	
L2	Puducherry Uttar Pradesh Gujarat Haryana Delhi Karnataka Andhra Pradesh	Karnataka Andhra Pradesh Tamil Nadu Kerala Uttar Pradesh West Bengal Goa	Tamil Nadu Kerala Delhi Haryana Gujarat Punjab Uttar Pradesh Madhya Pradesh West Bengal Assam	Puducherry	Karnataka Andhra Pradesh
L3	Rajasthan Chhattisgarh Himachal Pradesh Sikkim Uttaranchal West Bengal	Puducherry Himachal Pradesh Rajasthan Mizoram	Goa Orissa Puducherry Nagaland Tripura	Rajasthan Chhattisgarh Himachal Pradesh Sikkim Uttaranchal	West Bengal
L4	Assam Meghalaya Mizoram Orissa	Madhya Pradesh Assam Orissa Nagaland Chhattisgarh Jharkhand	Himachal Pradesh, Rajasthan, Chhattisgarh, Sikkim	Meghalaya Mizoram	Orissa, Assam
L5	Madhya Pradesh Bihar Jharkhand Lakshadweep Nagaland	Uttarakhand Tripura Sikkim Manipur Andaman and Nicobar Bihar Meghalaya	Jharkhand Andaman and Nicobar Uttarakhand Bihar Jammu and Kashmir Manipur Meghalaya	Lakshadweep	Madhya Pradesh Nagaland

(*continued*)

Table 4.5   (continued)

Levels	2005	2006	2008	Downward movement (from 2005 to 2008)	Upward movement (from 2005 to 2008)
L6	Andaman and Nicobar Arunachal Pradesh Dadra and Nagar Haveli Jammu and Kashmir Manipur Tripura	Lakshadweep Jammu and Kashmir Dadra and Nagar Haveli Arunachal Pradesh Daman and Diu	Lakshadweep Mizoram Arunachal Pradesh Daman and Diu Dadra and Nagar Haveli		Andaman and Nicobar Manipur Jammu and Kashmir Tripura

DIT & NCAER (2010)

Table 4.6   Readiness index for three years (2005, 2006 & 2008)

Levels	2005	2006	2008	Downward movement (from 2005 to 2008)	Upward movement (from 2005 to 2008)
L1	Andhra Pradesh Goa Punjab Maharashtra Tamil Nadu	Haryana Kerala Chandigarh	Haryana Chandigarh Karnataka Maharashtra Tamil Nadu	Andhra Pradesh Goa Punjab	
L2	Chandigarh Delhi Gujarat Haryana Karnataka Kerala Puducherry Uttar Pradesh	Karnataka Tamil Nadu Delhi Andhra Pradesh	Delhi Punjab Andhra Pradesh Kerala West Bengal Bihar	Gujarat Puducherry Uttar Pradesh	Chandigarh Haryana Karnataka

(continued)

Table 4.6   (continued)

Levels	2005	2006	2008	Downward movement (from 2005 to 2008)	Upward movement (from 2005 to 2008)
L3	Chhattisgarh Himachal Pradesh Rajasthan Sikkim Uttaranchal West Bengal	West Bengal Punjab Uttar Pradesh Maharashtra Goa Puducherry	Orissa Goa Madhya Pradesh Himachal Pradesh Gujarat Andaman and Nicobar	Rajasthan Chhattisgarh Uttaranchal Sikkim	West Bengal
L4	Assam Jharkhand Madhya Pradesh Meghalaya Mizoram Orissa	Gujarat Madhya Pradesh Himachal Pradesh Orissa Rajasthan, Lakshadweep Andaman and Nicobar Sikkim, Jharkhand Chhattisgarh Uttarakhand Meghalaya, Assam	Uttar Pradesh Rajasthan Chhattisgarh Jharkhand Sikkim, Uttarakhand	Assam Meghalaya Mizoram	Madhya Pradesh Orissa
L5	Bihar Jammu and Kashmir Lakshadweep Nagaland	Bihar Mizoram Jammu and Kashmir Nagaland Arunachal Pradesh Manipur Daman and Diu Dadra and Nagar Haveli	Assam Puducherry Tripura Jammu and Kashmir Meghalaya	Lakshadweep Nagaland	Bihar
L6	Daman and Diu Andaman and Nicobar Dadra and Nagar Haveli Manipur Arunachal Pradesh Tripura	Tripura	Nagaland, Mizoram Daman and Diu Manipur, Arunachal Pradesh Lakshadweep Dadra and Nagar Haveli		Tripura Andaman and Nicobar

The level of usage of ICT for governance in States/UTs has been shown in Table 4.7. The movement of the States/UTs across categories of the usage index has been presented for three years. The States and UTs that show relatively upward movement are Nagaland and Madhya Pradesh as they move up two levels. The regions with almost neutral or very less upward movement are Assam, Bihar, Jharkhand, Andaman and Nicobar and Uttaranchal. On the other hand, Haryana, Mizoram, Lakshadweep, Puducherry and Goa have shown significantly high negative growth. Other states have also shown some downward mobility including Kerala, Tamil Nadu, Maharashtra, Punjab, Meghalaya, Uttar Pradesh, Rajasthan, Gujarat, Andhra Pradesh, Sikkim and Uttaranchal, Assam, Meghalaya, Arunachal Pradesh, Daman and Diu and Jammu and Kashmir.

In terms of regions, eastern region has remained at same level while north-west and north-east region has witnessed a downward movement.

4.5   E-Governance Ranking Within Established Hierarchies (2008)

The E-Readiness report (2008) has developed a hierarchy on the basis of certain factors that indicate the successful implementation of E-Governance projects (Table 4.8). These factors include the following.

(a) An institutional mechanism such as a separate E-Governance department to implement E-Governance initiatives
(b) Web presence of E-Governance department and interactive portal with web links to implement the policies
(c) Strategic planning of E-Governance initiatives implementation
(d) Setting up of the State E-Governance Mission Team
(e) E-Governance budget for the particular department

Based on these indicators, three levels of hierarchy were finally proposed. Total nine States/UTs belong to the advanced (H1) category—Andhra Pradesh, Karnataka, Tamil Nadu, Punjab, Madhya Pradesh, Kerala, Chhattisgarh, Chandigarh and Gujarat. Eleven belong to middle level category of E-Governance: Jharkhand, Delhi, Haryana, Maharashtra, Goa, Rajasthan, West Bengal, Lakshadweep, Orissa, Uttar Pradesh and Uttarakhand and majority of them (15) lie in the primary category of E-Governance: Andaman and Nicobar, Sikkim, Meghalaya, Assam, Manipur, Himachal Pradesh, Daman and Diu, Tripura, Bihar, Dadra and Nagar Haveli, Arunachal Pradesh, Mizoram, Jammu and Kashmir, Puducherry and Nagaland.

Table 4.7 Usage index for three years (2005, 2006 & 2008)

Levels	2005	2006	2008	Downward movement (from 2005 to 2008)	Upward movement (from 2005 to 2008)
L1	Karnataka Chandigarh Delhi Kerala Haryana Tamil Nadu		Karnataka Chandigarh Andaman and Nicobar Delhi	Kerala Haryana Tamil Nadu	
L2	Goa Andhra Pradesh Maharashtra Mizoram Gujarat Punjab Rajasthan West Bengal	Chandigarh Chhattisgarh Delhi Karnataka	Andhra Pradesh Tamil Nadu West Bengal Kerala Jharkhand Gujarat Uttarakhand	Punjab Goa Mizoram Rajasthan Maharashtra	
L3	Uttar Pradesh Chhattisgarh Lakshadweep Meghalaya Sikkim Uttaranchal Himachal Pradesh	Rajasthan Maharashtra Himachal Pradesh Jharkhand Punjab Andhra Pradesh Tamil Nadu Mizoram Haryana Kerala West Bengal Gujarat Goa Uttar Pradesh	Haryana Punjab Sikkim Madhya Pradesh Maharashtra Rajasthan Himachal Pradesh Chhattisgarh Assam	Lakshadweep Uttar Pradesh Meghalaya	Uttaranchal

L4	Jharkhand Madhya Pradesh Orissa Puducherry	Sikkim Meghalaya Orissa Uttarakhand Assam Madhya Pradesh Andaman and Nicobar	Goa Uttar Pradesh Orissa Meghalaya Bihar	Puducherry	Jharkhand Madhya Pradesh
L5	Andaman and Nicobar Arunachal Pradesh Daman and Diu Jammu and Kashmir	Bihar Tripura Nagaland Daman and Diu Puducherry Manipur Arunachal Pradesh Lakshadweep	Mizoram Nagaland	Arunachal Pradesh Daman and Diu Jammu and Kashmir	Andaman and Nicobar
L6	Assam Bihar Dadra and Nagar Haveli Manipur Nagaland Tripura	Jammu and Kashmir Dadra and Nagar Haveli	Lakshadweep Dadra and Nagar Haveli Jammu and Kashmir Arunachal Pradesh Puducherry Daman and Diu Manipur Tripura		Assam Bihar Nagaland

DIT & NCAER (2010)

Table 4.8   E-Governance ranking within established hierarchies (2008)

Levels of hierarchy	Number of Indian states or Union Territories	Indian states or Union Territories (in order)
H1—Advanced	9	Andhra Pradesh
		Karnataka
		Tamil Nadu
		Punjab
		Madhya Pradesh
		Kerala
		Chhattisgarh
		Chandigarh
		Gujarat
H2—Middle	11	Jharkhand
		Delhi
		Haryana
		Maharashtra
		Goa
		Rajasthan
		West Bengal
		Lakshadweep
		Orissa
		Uttar Pradesh
		Uttarakhand
H3—Primary	15	Andaman and Nicobar
		Sikkim
		Meghalaya
		Assam
		Manipur
		Himachal Pradesh
		Daman and Diu
		Tripura
		Bihar
		Dadra and Nagar Haveli
		Arunachal Pradesh
		Mizoram
		Jammu and Kashmir
		Puducherry
		Nagaland

Source: DIT & NCAER (2010)

Most of the States and UTs classified under the 'advanced' category have qualified with all the relevant indicators like they have E-Governance-related promotion and implementation under separate institutional mechanism along with web presence and dynamic web portal linking to the services under E-Governance. They also have road maps to guide E-Governance initiatives and have set up States E-Governance Mission Team (SeMT). Moreover, a separate accounting and budgeting is done for the E-Governance-related projects and initiatives.

There are few exceptions like Karnataka and Chhattisgarh which are not having SeMT, while Tamil Nadu misses upon the implementation plan or road map. Areas like Chandigarh and Gujarat are not maintaining separate budgeting for the E-Governance initiatives. Due to lack of interactive web presence, states like Uttar Pradesh, Goa and West Bengal are not able to move to the advanced categorization.

In the 'medium' category of E-Governance, there is an existence of strategic implementation plan for E-Governance by all the States/UTs. But there are exceptions like Maharashtra, Delhi, Rajasthan and Orissa which are not having E-Governance-related separate institutional mechanism. Also, Maharashtra lags behind in both separate mechanism as well as budgeting. The state of Haryana lacks any dedicated task force or even a prominent web presence.

States and UTs in the 'primary' category lack one or more indicators. Absence of institutional mechanism, interactive portal or presence on web is prominent feature in this category. Manipur, Meghalaya, Mizoram and Nagaland lack separate institutional mechanism but does have an implementation plan and SeMT (Table 4.9).

Within a hierarchical category, ranking has been provided to the States/UTs to identify various factors for betterment. Principal Component Analysis (PCA) has been used for this purpose. Based on the analysis following conclusions have been made:

Karnataka has introduced less number of E-Governance projects with provision of citizen charters in less than 60 percent of the services. Karnataka needs to work upon proper implementation of its projects. The dissemination instruments such as radio, print, TV need to be improved in Delhi and Chandigarh. Moreover, there is only a ratio of 2:5 for projects of E-Governance per department in Delhi and Chandigarh. In case of Haryana, Punjab, Madhya Pradesh and Maharashtra, there are even lesser number of services with Citizen Charters. Low number of projects and no major road map for implementation was found in Uttar Pradesh. Similar is the case with the state of West Bengal, Jharkhand and Chhattisgarh with majority of the E-Governance initiatives toward people staying below poverty line. States like Andhra Pradesh, Tamil Nadu, Kerala, Madhya Pradesh, Rajasthan, Punjab, Gujarat, and Chhattisgarh are utilizing their E-Readiness status relatively well and have coordination within various

Table 4.9   Indicators of E-Governance ranking used for PCA

Major category	Minor category	Minor category indicators of significance
E-Governance	E-Governance projects	No. of E-Governance projects
	Age	Age of the oldest E-Governance project in the state
	BPR	Percentage of E-Governance projects with BPR
	Spread of E-Governance projects	Percentage of programs having entire state coverage as per plan
		Percentage of projects having infrastructure set up in entire state to total projects having entire state coverage as per plan
		Percentage of projects active in entire state to total projects having infrastructure set up in entire state
		Percentage of projects focusing on marginalized target groups
	Application in Services	Percentage of Services where changes have been implemented
		Percentage of Services which are transformation and transaction based
		Number of Services for which citizen charter provided
		Percentage of Services for which trips to the dealing office by user's availing the service has been completely eliminated
	Policy and Institutional Environment	Separate institutional mechanism to promote E-Governance
		Documented policy for E-Governance activities
		E-Governance road map document
		State E-Governance mission team (SeMT) set up for E-Governance projects
		Separate task force been set up for E-Governance projects
		Separate ICT budget
		Separate E-Governance budget
		Percentage of E-Governance projects to total number of departments
		Dissemination Instruments
		1. Radio
		2. Print
		3. Television
		4. Street Shows
		5. Any Other

Source: DIT & NCAER (2010)

government departments. These states have a higher E-Governance status. Tamil Nadu needs improvement in Citizen Charters for services. Kerala needs to focus on introduction and implementation on projects related to marginalized target groups.

4.6   E-GOVERNANCE INITIATIVES IN STATES

State governments have initiated E-Governance procedure in their particular states for IT implementation and online delivery of services to the citizens. The schemes introduced under the E-Governance initiative aim to provide Government to Citizen (G2C), Government to Business (G2B) and Government to Government (G2G) with the use of local languages. State governments have adopted specific Mission Mode Projects important for the economic development of the State. States, under Central Assistance, have introduced policies regarding Agriculture, Treasuries, Commercial Taxes, Road Transport, E-District, Employment exchange, Land Records, Police, Municipalities and Gram Panchayats.

4.6.1   Agriculture

Many State governments have introduced initiatives to provide web-based agricultural information and facilities to the public. It provides market price information to the farmers and traders so that they can receive competitive price for their agricultural produce. *KrishiMarataVahini* and *RaitaMitra* are two initiatives launched by Karnataka government related to agriculture for better cultivation and sale of agricultural produce. Online agriculture market price scheme has been launched in Meghalaya to disclose market trends related to agriculture price to the farmers. Farmer registration to receive advisories from experts on Agricultural issues is another service given by the government of Andaman and Nicobar, Arunachal Pradesh,[1] Bihar, Chhattisgarh and Assam, Tripura, Manipur, Meghalaya, Mizoram, Madhya Pradesh, Nagaland, Puducherry and Telangana. Farmer registration and certification is done in Andhra Pradesh.[2] Governments in Assam, Bihar, Himachal Pradesh, Haryana, Tamil Nadu, Sikkim, Jammu and Kashmir,[3] Karnataka, Maharashtra,

---

[1] http://www.andaman.gov.in/web/guest/andaman-home. Accessed on 15th December 2017.

[2] http://apnacsctest.co.in/. Accessed on 15th December 2017.

[3] https://sarathi.nic.in:8443/nrportal/sarathi/HomePage.js.   Accessed   on   15th December 2017.

Meghalaya, Manipur, Madhya Pradesh, Nagaland, Odisha, Punjab, Tamil Nadu, Rajasthan, Uttarakhand, Uttar Pradesh and West Bengal have introduced facility of agriculture weather insurance to get the farmers insured for their crops against unfavorable weather conditions.[4] The Himachal Pradesh government provides agriculturist certificate that certifies the applicant as a farmer or an agriculturist. This certification can be availed through E-Governance service. The facility of *AgriSubsidy* application for the farmers to receive agriculture subsidies on various central and state government schemes is provided by Chhattisgarh.[5] Agriculture income certificates used for getting bank loans are issued and Crop insurance scheme to help the farmers against the financial losses by the government of Andhra Pradesh, Gujarat,[6] Maharashtra and Telangana.

State governments need to introduce similar schemes to facilitate better cultivation and sale of agricultural produce and to help farmers by providing them information about the current market price. Also, information about recent technologies that can be used to improve the agricultural produce also needs to be provided through E-Services by the State governments.

Khasra pahani is a legal agricultural document used in Telangana that specifies and keeps records of land and crop details and is used to know the genuineness of seller (owner) when land is being purchased. Using this service, user can get the certified copy of *khasra pahani* to show at Sub-Register's office when sale transaction is being done.

4.6.2   Commerce and Industry

Manufacturer license issuance, renewal and market license issuance for establishment of shops through Municipal Corporations and Councils is done through E-Governance service in Maharashtra and Chhattisgarh.

4.6.3   Tourism

Hotel bookings for the people visiting the city for its culture and social existence are provided by the Tourism development department of the State. This facility is available in Kerala, Maharashtra[7] and Madhya Pradesh.

---

[4] http://csc.gov.in/states/assam/. Accessed on 15th December 2017.
[5] http://cg.nic.in/agrisubsidy. Accessed on 15th December 2017.
[6] http://portal.gujarat.gov.in/#. Accessed on 15th December 2017.
[7] https://www.mahaonline.gov.in/Molweb/Site/Home/Index.aspx. Accessed on 15th December 2017.

4.6.4   Consumer Affairs, Food and Public Distribution

The State governments of Punjab, Karnataka and Assam have launched E-Governance schemes related to consumer affairs and public distribution system. *Ahara* (Karnataka), *GRIHA-LAKSHMI Computerized Public Distribution System (PDS)* (Assam), *Punjab Sewa* Online and *E-District* (Punjab) are public distribution system schemes that provide online information to BPL families regarding registration, retail prices, district-wise allotment of food grains, change of address, addition of members, issue of duplicates and so on. These schemes of public distribution facilitate provision of information on essential public commodities like food grains, sugar, petroleum products and generate various reports which help in decision making, that is, ration card details, fair price shops details, and oil depot details. Similar schemes have been introduced in Andaman and Nicobar,[8] Maharashtra, Manipur, Madhya Pradesh, Telangana, Uttar Pradesh, Mizoram and Rajasthan. Policies related to consumer affairs, food and public distribution system needs to be introduced in various states to facilitate dissemination of information about public distribution to BPL families and provide them online facilities to issue a new ration card, make changes related to address or members or get details of fair price shops. Chhattisgarh,[9] Andaman and Nicobar, Dadra and Nagar Haveli,[10] Karnataka, Tripura, Gujarat,[11] Kerala, Manipur, Nagaland, Punjab and Sikkim facilitate issuance of digital ration card. Gujarat government enables *PDS SMS alerts* and *PDS transit passes* printed. Chhattisgarh government provides *PDS truck dispatch* information through SMS services, Smart Ration Card to BPL beneficiaries, FPS automation facility and Centralized Online Real Time Electronic Public Distribution System to improve service delivery. Maharashtra government has introduced Grievance Monitoring and Complaint Redressal System for Public Distribution System (*Takrar*) where users can complain through a toll-free number.[12]

Telangana government has introduced the service that facilitates food security cards to citizens. By presenting the food security cards at MRO office, citizens can get their ration.

---

[8] http://www.andaman.gov.in/web/guest/andaman-home. Accessed on 15th December 2017.

[9] https://edistrict.cgstate.gov.in/PACE/login.do. Accessed on 15th December 2017.

[10] http://pdsportal.nic.in/. Accessed on 15th December 2017.

[11] https://patan.gujarat.gov.in/home. Accessed on 15th December 2017.

[12] http://mahafood.gov.in. Accessed on 15th December 2017.

4.6.5   Railways

Online train ticket booking, enquiry and cancellation at IRCTC are services provided by many State Governments such as Arunachal Pradesh, Assam, Bihar, Daman and Diu, Tamil Nadu, Delhi, Dadra and Nagar Haveli, Gujarat, Himachal Pradesh, Jammu and Kashmir, Karnataka, Telangana, Kerala, Mizoram Lakshadweep, Meghalaya, Nagaland, Odisha, Punjab, Puducherry, Rajasthan, Uttarakhand, West Bengal and Uttar Pradesh. Booking as IRCTC agent is facilitated in Gujarat, Himachal Pradesh, Jammu and Kashmir,[13] Karnataka, Kerala, Puducherry, Lakshadweep, Maharashtra, Tamil Nadu, Meghalaya, Manipur, Madhya Pradesh, Nagaland, Odisha, Punjab, Sikkim, Telangana, Tripura, Uttar Pradesh and West Bengal.

4.6.6   Labor

Employment Information Service provides information of latest openings in government departments. This service is provided by various state governments like Andhra Pradesh and Assam. In Maharashtra, candidates can enroll for ITI course and job seekers can also register in government employment exchange using this service. *Employment Information Service* initiated in Telangana where latest openings in Telangana State government are made available to the job seekers.

4.6.7   BFSI

Application of case listing is a facility provided by many States that belong to the Revenue Department. It allows citizens to apply for case listing through *Lok Seva Kendra* or through department counters. The government of Chhattisgarh provides this facility.[14]

Khajane project in Karnataka involves online treasury computerization system that has the ability to track every activity from approval of State government to rendering accounts to the government.

General insurance policy records are maintained through this service in many States such as Kerala, Andaman and Nicobar, Andhra Pradesh, Chhattisgarh, Assam, Gujarat, Himachal Pradesh, Maharashtra,

---

[13] http://www.apna.csc.gov.in/index.php/. Accessed on 15th December 2017.
[14] https://edistrict.cgstate.gov.in/PACE/login.do. Accessed on 15th December 2017.

Meghalaya, Manipur, Madhya Pradesh, Mizoram, Uttar Pradesh, Tamil Nadu, Odisha, Punjab, Puducherry, Rajasthan, Telangana and Tripura, the facility of collection of LIC premium at their respective centers.

Service of online pan card application like form filling, attaching attested certificates and IDs is provided by Odisha, Punjab, Puducherry, Sikkim, Telangana and Uttar Pradesh government.

4.6.8   Health and Family Welfare

Telemedicine facility is being provided by many States to bring healthcare within reach of rural population, to provide help in medical cases, to provide specialized help in remote areas and for the doctors to discuss with their peers and handle complicated cases. Chhattisgarh, Assam, Bihar, Delhi, Gujarat, Himachal Pradesh, Haryana, Jharkhand, Jammu and Kashmir, Kerala, Maharashtra, Madhya Pradesh, Odisha, Punjab, Puducherry, Rajasthan, Tamil Nadu, Tripura, Uttarakhand, Uttar Pradesh and West Bengal.

Aam Aadmi Bima Yojana (*AABY*).

Medical Insurance, online registration of number of patients admitted to the hospitals is the health service introduced in Andhra Pradesh. Rajiv *Arogyasri* is the flagship health initiative launched in Andhra Pradesh under which all health initiatives are introduced to provide quality healthcare to the poor. Online health services are provided in Meghalaya.

4.6.9   Human Resource Development

Online open school societies, teacher eligibility tests, District Institutes of Education and Training have been established by many State governments. *CSC academies* have been introduced in many States to provide training to rural masses in Delhi, Himachal Pradesh, and Madhya Pradesh. Ancient *Gurukul* learning, a medium to exchange practical and theoretical knowledge of arts by experts, is made possible online by using this service in Jharkhand, Uttarakhand, Jammu and Kashmir, Karnataka, Kerala, Maharashtra, Meghalaya, Madhya Pradesh, Odisha, Punjab, Puducherry, Tamil Nadu, Tripura, Rajasthan and Uttar Pradesh. *Annual works* and *Finance Plan Indira KranthiPantham* is the skill development program launched in Andhra Pradesh[15] for the poor and the youth for placement in

---

[15] http://www.esevaonline.ap.gov.in. Accessed on 15th December 2017.

construction, textile and service industry. Programs on job skill develop-
ment have also been announced in Assam,[16] Himachal Pradesh, Jharkhand,
Jammu and Kashmir, Kerala, Puducherry, Maharashtra, Madhya Pradesh,
Tripura, Uttar Pradesh and West Bengal. *E-Kalyan* scheme launched in
Jharkhand[17] provide information to scholarship applicants (students)
within or outside the State through SMS. Admission fee can also be paid
online through this service to get admission in government colleges.
Kerala, Delhi, Maharashtra and Gujarat provide this service. The facility of
online display of Classes 10 and 12 results is provided by Madhya Pradesh[18]
government.

To promote girl education, Ladli Lakshmi scheme has been launched in
Madhya Pradesh that provides scholarship to girls. Marriage Assistance
Plan[19] has also been introduced to provide money to parents who cannot
afford money for the marriage of their children. Similarly, *EVR
Maniammaiyar Ninaivu Widow Daughter Marriage Assistance Scheme* has
been introduced in Tamil Nadu to provide financial assistance to poor
status so that they can educate and marry their daughters. Uttar Pradesh
government has also launched a scheme to provide grant to the destitute
widows for the marriage of their daughters.

4.6.10   *Information, Technology and Broadcasting*

Many State governments have provided the service of *Wallet TopUp*. This
service fill-up the amount via debit card/credit card and internet banking
and this amount will be transferred by any registered mobile instantly.
Delhi, Gujarat, Goa, Himachal Pradesh, Haryana, Jharkhand, Jammu and
Kashmir,[20] Kerala, Maharashtra, Meghalaya, Manipur, Tripura, Madhya
Pradesh, Nagaland, Odisha, Punjab, Puducherry, Tamil Nadu,
Uttarakhand, Uttar Pradesh and West Bengal.

There is a facility of *Adhaar* Card being printed, status can be generated, and Adhaar card can be printed from the website. *Unique Identification
number (UID)* assigned to all citizens can also be obtained from the website. States that provide this facility are Kerala, Delhi, Lakshadweep,

---

[16] http://csc.gov.in/states/assam/. Accessed on 15th December 2017.
[17] http://ekalyan.jharkhand.gov.in/. Accessed on 15th December 2017.
[18] http://mpbse.nic.in/results.htm. Accessed on 15th December 2017.
[19] http://mpedistrict.gov.in. Accessed on 15th December 2017.
[20] http://apna.csc.gov.in/. Accessed on 15th December 2017.

Gujarat, Goa, Himachal Pradesh, Haryana, Jharkhand, Tripura, Jammu and Kashmir, Maharashtra, Uttarakhand, Uttar Pradesh, West Bengal, Mizoram, Manipur, Meghalaya, Madhya Pradesh, Mizoram, Nagaland, Odisha, Punjab, Puducherry, Sikkim, Telangana and Tamil Nadu.

The facility of Adhaar-E-KYC is provided by many State governments such as Manipur, Madhya Pradesh, Nagaland, Odisha, Punjab, Puducherry, Rajasthan, Sikkim, Tripura, Tamil Nadu, Uttarakhand, Uttar Pradesh and West Bengal. Aadhaar-E-KYC is an electronic record provided by UIDAI. The customer and Village Level Entrepreneur input his/her biometrics data via a UIDAI biometric reader. The UIDAI then transfers the data of the individual comprising name, age, gender and photograph of the individual electronically to the bank/BCs without actual submission of the physical copy of address proof and photo.

Service of Beam Telecom Payment is provided by Telangana government. The payment for Beam fiber, the leading internet broadband service providers in Hyderabad, can be made through E-Seva centers located in many areas of the State.

4.6.11   *Law and Justice*

Online applications for property tax payment, passport issuance, change of passport details, issue of new visa, visa conversion, water bill payment, water tax payment and RTI filling at municipalities/panchayat have been introduced in Chhattisgarh, Puducherry, Andhra Pradesh,[21] Delhi,[22] Arunachal Pradesh, Chandigarh,[23] Delhi,[24]Gujarat[25,26] Himachal Pradesh, Tamil Nadu, Sikkim, Haryana,[27] Jammu and Kashmir, Karnataka, Kerala,[28] Uttarakhand, Lakshadweep, Maharashtra, Madhya Pradesh, Nagaland, Odisha, Punjab, Puducherry, Tripura, Uttarakhand, Uttar Pradesh and West Bengal. *E-CST module* has been designed for sales tax payment in Daman and Diu and Dadra and Nagar Haveli.

[21] http://www.esevaonline.ap.gov.in. Accessed on 15th December 2017.

[22] http://indianfrro.gov.in/frro/menufrro.jsp. Accessed on 15th December 2017.

[23] http://164.100.147.15/epayment/StaticPages/elec_water_bill.aspx. Accessed on 15th December 2017.

[24] http://esla.delhi.gov.in/. Accessed on 15th December 2017.

[25] http://indianfrro.gov.in/frro/menufrro.jsp?t4g=QAO45XTW. Accessed on 15th December 2017.

[26] https://www.suratmunicipal.gov.in/epay/?SrNo=405005305405. Accessed on 15th December 2017.

[27] http://indianfrro.gov.in/frro/menufrro.jsp?t4g=QAO45XTW. Accessed on 15th December 2017.

[28] https://edistrict.kerala.gov.in/. Accessed on 15th December 2017.

Nagaland government provides VAT registration facility online. People are not required to visit the VAT office; VAT, C forms and other statutory compliances can be fulfilled through department website.[29]

Odisha government provides electronic welfare schemes.[30] The database is digitized and service delivery is also through electronic means.

Punjab government provides online right to information[31] service to ensure service delivery for requested information under the RTI Act. Using the E-Governance service, Telangana government has facilitated request for new gas connection that can be made by the customers online or by manually applying to the place.

Service of *Centralized Treasury System E-Payment* is facilitated in Uttar Pradesh where all Treasury transactions are done through net banking E-Payment. All treasury transactions are credited directly into Employee/ Pensioner or third party account without involvement of cash or check.

FRIENDS[32] (Fast, Reliable, Instant, Efficient Network for the Disbursement of Services) is a Single-Window Facility introduced by Kerala government to provide citizens the means to pay taxes and other financial dues to the Government.

4.6.12   *Power*

The facility of online electricity bill payment through application by consumers to the State Power Distribution is provided by Chhattisgarh, Andaman and Nicobar,[33] Chandigarh,[34] Delhi,[35] Dadra and Nagar Haveli, Gujarat,[36] Himachal Pradesh, Haryana, Jharkhand, Karnataka, Kerala,[37] Lakshadweep, Maharashtra, Meghalaya, Manipur, Madhya Pradesh, Odisha, Punjab, Puducherry, Rajasthan, Telangana, Tripura, Uttarakhand, Uttar Pradesh and West Bengal.

---

[29] http://taxsoft-ngl.nic.in/nagalandereg/RegistrationOptions.aspx. Accessed on 15th December 2017.

[30] https://www.ulbodisha.gov.in/or/emun/home. Accessed on 15th December 2017.

[31] http://www.punjab.gov.in/departmental/. Accessed on 15th December 2017.

[32] http://www.itmission.kerala.gov.in/friends.php. Accessed on 15th December 2017.

[33] http://www.andaman.gov.in/web/guest/andaman-home. Accessed on 15th December 2017.

[34] http://164.100.147.15/epayment/StaticPages/elec_water_bill.aspx. Accessed on 15th December 2017.

[35] http://www.tatapower-ddl.com/cmspage.aspx?section=Bill   Payment&tabname=Pay Bill Online&level=1. Accessed on 15th December 2017.

[36] http://swagat.gujarat.gov.in/. Accessed on 15th December 2017.

[37] https://edistrict.kerala.gov.in/FREES. Accessed on 15th December 2017.

Telangana government provides a 25 percent subsidy on power bills for a period of years or the amount of bill whichever is earlier from the date of commercial operations can be done through this service.[38] They also provide new electricity connection through this service. The service is available in religious places and government schools of Adilabad, Khammam, Karimnagar, Nizamabad and Warangal Districts.

Centralized Utility Approval System (CUAS) is an E-Governance initiative in Uttarakhand Power Corporation Ltd. (UPCL) to facilitate new electricity connections to its consumers in a transparent manner.

4.6.13   Personnel, Public Grievances, Pension

Various State governments have launched E-Governance schemes related to public grievances and pensions. The government of Assam,[39] Bihar, Haryana, Tamil Nadu, Jharkhand, Jammu and Kashmir,[40] Karnataka, Maharashtra, Manipur, Madhya Pradesh, Tripura, Odisha, Punjab, Puducherry, Rajasthan, Uttarakhand, Uttar Pradesh and West Bengal has launched a speedy redressal system and effective monitoring of the grievances besides providing a fast access to the pensioners. Himachal Pradesh[41] government's electronic grievance redressal system provides facilities like service request, approval of request, database maintenance and delivery. It provides the number of complaints on which some action has been taken. *Harsamadhan* portal of Haryana is the redressal system that contains complete workflow and reporting mechanism.

Indira Gandhi Old Age Pension scheme has been launched in Chhattisgarh and Madhya Pradesh to provide financial assistance to old age infirm persons.[42] The facility of grievance redressal through meeting with the chief minister is available in Chhattisgarh. *SWAGAT*[43] (State-wide Attention on Public Grievances by Application of Technology) is an E-Governance initiative launched in Gujarat that enables direct communication between the citizens and the chief minister for grievance redressal.

---

[38] http://www.meeseva.gov.in/Meeseva/intro.html. Accessed on 15th December 2017.

[39] http://csc.gov.in/states/assam/. Accessed on 15th December 2017.

[40] http://www.apna.csc.gov.in/index.php/. Accessed on 15th December 2017.

[41] http://esamadhan.nic.in. Accessed on 15th December 2017.

[42] https://edistrict.cgstate.gov.in/PACE/login.do. Accessed on 15th December 2017.

[43] http://gil.gujarat.gov.in/swagat.html. Accessed on 15th December 2017.

Governments of many States have launched *PFRDA* which is a national pension system to provide adequate solution to the problem of retirement income such as Andaman and Nicobar,[44] Arunachal Pradesh,[45] Assam,[46] Delhi, Gujarat,[47] Odisha, Tripura, Himachal Pradesh, Haryana, Karnataka, Puducherry, Punjab, Telangana, Tamil Nadu, Jammu and Kashmir, Jharkhand, Maharashtra, Meghalaya, Manipur, Nagaland, Madhya Pradesh and Mizoram. *Social service schemes*[48] and *social welfare payments*[49] have been introduced in Haryana. Social service schemes involve direct disbursement of pensions such as old age widow while social welfare payments prepare a list of beneficiaries and make electronic payment into account of stakeholders. Old age widow pensions have also been introduced in Karnataka, Kerala,[50] Punjab, Uttarakhand, Uttar Pradesh and West Bengal. Ministry of Social Justice and Empowerment is responsible for welfare of the Senior Citizens in Maharashtra.[51] Old age widow pension scheme has also been introduced in Manipur, Madhya Pradesh and Mizoram. *Anuprati*[52] scheme has been launched in Rajasthan for social welfare through internet or E-Mitra/CSC outlet.

Prajavani[53] is a grievance cell launched in Telangana for people to apply directly to government departments from MeeSeva centers in case of any issue.

Telangana government has initiated a service which deals with implementation of *Aasara Social Security Pension Scheme*[54] for old infirm widows, differently abled weavers and Toddy Tappers with HIV-AIDS. They also launched Pension and Insurance scheme for the SHG women to provide monetary help in their old age with the name of *Abhayahastam*.[55]

---

[44] http://www.andaman.gov.in/web/guest/andaman-home. Accessed on 15th December 2017.

[45] csc.gov.in/states/ap/. Accessed on 15th December 2017.

[46] http://csc.gov.in/states/assam/. Accessed on 15th December 2017.

[47] http://swagat.gujarat.gov.in/. Accessed on 15th December 2017.

[48] http://socialjusticehry.gov.in. Accessed on 15th December 2017.

[49] http://socialjusticehry.gov.in. Accessed on 15th December 2017.

[50] https://edistrict.kerala.gov.in. Accessed on 15th December 2017.

[51] https://www.mahaonline.gov.in/Molweb/Site/Home/Index.aspx. Accessed on 15th December 2017.

[52] http://rajpms.nic.in/. Accessed on 15th December 2017.

[53] http://www.nrcfoss.org.in/repository/e-gov/e-governance/prajavani. Accessed on 15th December 2017.

[54] http://www.esevaonline.telangana.gov.in/htmlpages/login.htm. Accessed on 15th December 2017.

[55] http://www.esevaonline.telangana.gov.in/htmlpages/login.htm. Accessed on 15th December 2017.

4.6.14 Shipping Road Transport and Highways

New learner license, new vehicle registration, driving license issuance, vehicle challan payment, vehicle transfer and conversion of vehicle from private to commercial and from commercial to private are transport services provided by the government of Chhattisgarh, Andaman and Nicobar, Arunachal Pradesh, Assam, Bihar, Chandigarh, Daman and Diu, Delhi,[56] Meghalaya, Dadra and Nagar Haveli, Goa, Gujarat, Himachal Pradesh, Haryana, Jharkhand, Jammu & Kashmir, Karnataka, Kerala,[57] Lakshadweep, Maharashtra, Tripura, Manipur, Madhya Pradesh, Mizoram, Nagaland, Odisha, Punjab, Puducherry, Rajasthan, Sikkim, Telangana. Online traffic challan payment is feasible in Telangana, Tamil Nadu, Uttarakhand, Uttar Pradesh and West Bengal.

4.6.15 Social Justice and Empowerment

Birth Certificate, Death Certificate, Marriage Certificate, and OBC/ST/ SC Certificate issuance, pregnant women registration through online modes is facilitated by governments in Chhattisgarh, Andaman and Nicobar,[58] Arunachal Pradesh, Assam, Bihar, Dadra and Nagar Haveli,[59] Gujarat,[60] Himachal Pradesh,[61] Haryana,[62] Jharkhand,[63] Tripura, Telangana, Uttarakhand, Uttar Pradesh, West Bengal, Mizoram, Nagaland, Jammu and Kashmir, Karnataka,[64] Kerala, Tamil Nadu, Lakshadweep, Maharashtra, Manipur, Meghalaya,[65] Madhya Pradesh,[66] Mizoram, Nagaland, Odisha, Punjab,[67] Puducherry, Uttar Pradesh and Sikkim.

---

[56] https://vahan.nic.in/. Accessed on 15th December 2017.

[57] https://edistrict.kerala.gov.in. Accessed on 15th December 2017.

[58] http://www.andaman.gov.in/web/guest/andaman-home. Accessed on 15th December 2017.

[59] http://nrhm-mcts.nic.in. Accessed on 15th December 2017.

[60] http://ahmedabadcity.gov.in/portal/index.jsp. Accessed on 15th December 2017.

[61] http://admis.hp.nic.in/epraman. Accessed on 15th December 2017.

[62] http://edisha.gov.in. Accessed on 15th December 2017.

[63] https://serviceonline.gov.in/. Accessed on 15th December 2017.

[64] http://nadakacheri.karnataka.gov.in. Accessed on 15th December 2017.

[65] https://meghalayaonline.gov.in/certisoft3/citizensscst/scstapplicationform.jsp. Accessed on 15th December 2017.

[66] http://mpedistrict.gov.in. Accessed on 15th December 2017.

[67] http://edistrict.punjab.gov.in/EDA/index.aspx?ActorTypeID=2. Accessed on 15th December 2017.

Correction of Birth, Death and Marriage Certificate is also feasible through these online services in many states as Himachal Pradesh, Delhi,[68] Haryana, Odisha and Telangana government provide unemployment allowance to people who are unemployed and disabled or have special needs to help ensure their survival through this service. Through this service, a bonafide Himachali Certificate can also be obtained.

Puducherry government provides *E-Nargrik service* that helps facilitate certificates like caste/residential/income certificate. Telangana government has introduced *Apathbandhu*[69] Scheme that provides accident insurance for accidental deaths to the families below poverty line.

The government of Telangana state, through *MeeSeva* service, provides monetary help/scholarship to the scheduled caste, scheduled tribe, backward class, disabled and economically backward class students. The Telangana government also launched financial plan *Aajeevika*[70] to alleviate rural poverty, promote sustainable livelihoods, social empowerment and rural development. Finance Plan *Mahila kishan sasakthikaran pariyojana* is another initiative by Telangana government to promote rural development and carry out welfare and integrated programs for the development of poor and needy, especially in rural areas. *Finance Plan—NON HRMS SALARIES* also work for the rural poor.

Damppati Puraskar Scheme is a scheme launched in Uttar Pradesh for citizens to apply for grant for re-marriage of widow.

4.6.16   Urban Development

Land allotment, its electronic submission and related information storage in the depository database of CSIDC department is the service provided by many State governments such as Chhattisgarh, Bihar, Karnataka, and Telangana. Facility of online land records has been maintained in Daman and Diu, Dadra and Nagar Haveli, Gujarat,[71] Tripura, Telangana, Himachal Pradesh[72] and Maharashtra[73] to provide details of document and the personal details of the citizen buying or selling. Application of

---

[68] http://mission.delhi.gov.in/. Accessed on 15th December 2017.

[69] http://www.it.telangana.gov.in/e-governance/. Accessed on 15th December 2017.

[70] http://www.esevaonline.ap.gov.in. Accessed on 15th December 2017.

[71] https://anyror.gujarat.gov.in/. Accessed on 15th December 2017.

[72] http://admis.hp.nic.in/himbhoomilmk. Accessed on 15th December 2017.

[73] https://mahabhulekh.maharashtra.gov.in/. Accessed on 15th December 2017.

property registration is the facility provided by Gujarat government to make the related information easily available. Land record of particular owner is maintained using halris/haris software at all tehsils across Haryana using this service. *Nakal of ROR* is another service in Haryana like land record maintenance except that it maintains revenue records. *Bhoomi* project in Karnataka is a self-sustainable E-Governance project for online delivery of land records. *Lokvani* project in Uttar Pradesh is another project initiated for land records delivery online.

CARD[74] (Computer-Aided Administration of Registration Department) is a project launched in Andhra Pradesh that facilitates complete computerization of land registration and records.

Indira Awass Yojana[75] is a housing scheme launched by the Tamil Nadu government which aims to provide housing facilities to the families which are below poverty line. The progress of construction and payment to the beneficiary is done digitally.

The State governments have launched E-Governance initiatives in various sectors—urban development, social justice and empowerment, transport, public grievance, pension, law and justice, power, railways, Information, Technology and Broadcasting, Human Resource Development, Health and Family welfare, Banking and Finance, Rural Development, labor, tourism, commerce and industry, and agriculture. The E-Governance initiatives of government aim to build an informed society, increase government and citizen interaction, enhance transparency and accountability in governance process and provide public services to the citizens and businesses.

Private sector has already launched services for the public in various sectors. Wallet TopUp, online electricity bill payment facility, mobile bill payment, medical insurance, general insurance, hotel bookings and other services have already been introduced by the private sector. The public sector is, however, novice to these services and has come up with similar facilities to match up the level of services provided by the private sector. The government is making efforts to provide facilities digitally and become more economically competitive to catch up with the private sector (Table 4.10).

A list of state-wise E-Governance initiatives is shown in Appendix.

---

[74] http://www.nic.in/projects/computer-aided-administration-registration-department-card. Accessed on 15th December 2017.

[75] http://www.tnrd.gov.in/. Accessed on 15th December 2017.

Table 4.10 E-Participation index of states

	State name	Online availability & Presence Index (OI)	Telecom Index (TII)	Human Capital Index (HCI)	Infrastructure Index (II)	E-Participation Index (EPI)	EGDII
1.	Andhra Pradesh	0.658294731	0.373680748	0.843151751	0.230574394	0.209930812	0.463126587
2.	Arunachal Pradesh	0.55605609	0.084405519	0.642377366	0.423590573	0	0.34128593
3.	Assam	0.861634947	0.028069094	0.161786099	0.21005117	0.010150897	0.254338442
4.	Bihar	0.900142533	0	0	0.091856972	0.014770673	0.201354036
5.	Chhattisgarh	0.696629058	0.070579087	0.358025493	0.191306827	0.098977609	0.283103615
6.	Goa	0.135956251	0.247140709	0.524563261	0.123813388	0.048710415	0.216036805
7.	Gujarat	0.768928626	0.218476034	0.466371665	0.421610988	1	0.575077463
8.	Haryana	0.795544427	0.162415403	0.444843222	0	0.159856306	0.312531872
9.	Himachal Pradesh	0.908972676	0.301935451	0.66617146	1	0.057889967	0.586993911
10.	Jammu and Kashmir	0.946268527	0.103596556	0.261335689	0.130354736	0.007967291	0.28990456
11.	Jharkhand	0.720144962	0.008865218	0.240155534	0.208487863	0.037781004	0.243086916
12.	Karnataka	0.816816841	0.282451508	0.399057596	0.051789653	0.228811656	0.355785451
13.	Kerala	1	0.583584844	0.607661815	0.050839284	0.162535937	0.480924376
14.	Madhya Pradesh	0.693252202	0.070777912	0.454333026	0.258998107	0.772057071	0.449883664

15. Maharashtra	0.918876236	0.341086021	0.540636872	0.309607714	0.09692143	0.441425655
16 Manipur	0.037907558	0.078562264	0.849004442	0.425247885	0.015028146	0.281150059
17. Meghalaya	0	0.073153705	0.6951787	0.251000567	0.134034706	0.230673536
18. Mizoram	0.760875916	0.086723658	1	0.5500582	0.038720876	0.48727573
19. Nagaland	0.051756689	0.080598213	0.363683272	0.325332616	0.041133941	0.172500946
20. Odisha	0.812677676	0.05886603	0.287183515	0.293366889	0.046058257	0.299630473
21. Punjab	0.555546432	0.339225557	0.327782598	0.112680982	0.432798905	0.353606895
22. Rajasthan	0.770638494	0.116019379	0.259271407	0.146099626	0.094643842	0.277334549
23. Sikkim	0.610456686	0.102257441	0.765612642	0.258581922	0.000828675	0.347547473

Source: Gupta et al. (2016a, b, c, d, e, f, g, h)

4.7   EGDII

4.7.1   States

The E-Governance Development Index for India (EGDII) of the states shows the E-Governance development among Indian states for the year 2014–2015. This was a pilot study done by the same set of authors (Gupta et al. 2016a, b, c, d, e, f, g, h). The enquiry tells E-Governance performance on the basis of the indicators—Online Availability and Presence, Telecom, Human Capital, Infrastructure and E-Participation. The analysis reveals that the southern zone is the leader in E-Governance services and its implementation.

The southern zone is the pioneer in introducing the concept of E-Governance services in India and has the best supporting infrastructure that provides the reason for its success. The highest literacy level in southern states provides support to the high Human Capital Indices in the zone, the strong telecommunication facilities and online portals and services quite support its success in implementation of E-Governance policies despite limited number of services and Common Services Centres. In southern zone, Tamil Nadu and Andhra Pradesh have a strong EGDII and have shown consistent performance in all indices.

The second-best performance has been shown by the western zone where the states of Gujarat and Maharashtra have performed quite well leading in online presence, telecommunication services and infrastructure-supporting E-Governance.

Behind these zones, third position is taken by north central zone where Uttarakhand and Madhya Pradesh have well-established and implemented E-Governance policies. The least performing zone is eastern zone.

4.7.1.1   Leading States

The leader states/UTs include Himachal Pradesh, Gujarat, Uttarakhand, Tamil Nadu, Mizoram, Lakshadweep, Delhi and Chandigarh. Himachal Pradesh has got the highest EGDII due to the first rank in Infrastructure Index (high number of Common Service Centres and SWAN networks) and exceptionally high rating in the Online Presence Index. The State has implemented various E-Governance schemes successfully such as E-Samadhan, E-Gazette, telemedicine, CSC academies and others.

The next highest ranking is of Gujarat due its highest E-Participation index and high online presence and infrastructural facilities. Its successful E-Governance projects include Mahiti Shakti, E-Dhara, Jan Sewa Kendra, SWAGAT and E-Gram Viswa Gram Project.

The next best performer is Uttarakhand. The State has highest Telecom Index, high Online Availability and Performance Index and high Human Capital Index. For the welfare of its citizens, various schemes have been launched—Devbhoomi (Uttarakhand Land Records), Online Content Creation/IT-Enabled Course Curriculum, School Education Portal and Centralized Utility Approval System (CUAS).

Tamil Nadu is placed at the fourth position. The state has commendable E-Participation and online presence. The state runs a large number of welfare schemes such as EVR Maniammaiyar Ninaivu Widow Daughter Marriage Assistance Scheme, Dr. Dharmambal Ammaiyar Ninaivu Widow Re-Marriage Assistance Scheme, online land records and employment online.

Although the eastern zone overall has performed low, Mizoram has unexpectedly shown high performance which is attributed to high Human Capital Index. Mizoram has the highest human capital index due to high literacy rate, significant youth population and an above average Gross Enrollment Ratio. The State government runs successful education schemes such as K-Yan.[76]

4.7.1.2 Potential Leaders

Due to its high ranking in Online Availability and Infrastructure Index, Kerala ranks first in potential leaders. The State has provided adequate infrastructural facilities and has good web presence for assistance to its citizens. Human Capita Index is also high for the State due to its high literacy rate. The State government has initiated many policies to improve its status of E-Governance such as Akshaya, BhuRekha and FRIENDS.

Andhra Pradesh takes second position in the EGDII index due to its high online presence and Human Capital Index. The State portal is active but citizen interaction is lacking. E-Governance efforts have been made to improve Telecom, Infrastructure and E-Participation Index such as MeeSeva, Complaint Redressal System, Employee Information System (EIM)-Department of School Education, Prajavani (AP)—an E-Effort to Empower, *AABY*—AamAadmiBimaYojana Medical Insurance, Arogyasri, *Annual works* and *Finance Plan Indira KranthiPantham* and *CARD*[77] (Computer-Aided Administration of Registration Department).

---

[76] https://msegs.mizoram.gov.in/. Accessed on 15th December 2017.

[77] http://www.nic.in/projects/computer-aided-administration-registration-department-card. Accessed on 15th December 2017.

Madhya Pradesh and Maharashtra are the next states in the list of potential leaders with significantly high online availability. The online portals of both the states have adopted local language for easy access and facilitate high user interactivity. Madhya Pradesh has a strong E-Participation index which Maharashtra lacks. Both the states need to improvise on their Telecom and Infrastructure Index.

The last state in the list of potential leaders is Karnataka with high web presence. The web portal is highly interactive, but the infrastructure does not support the policy implementation which poses as an obstruction for the State to become a leader from potential leader. The state has launched a lot of E-Governance initiatives such as Bhoomi, SamanyaMahiti, E-Granthalaya, SahakaraDarpana of Directorate of Co-operative Audit, Sarathi and Vahan of Transport Department, Returns Filing System (RFS), E-MAN, AasthiTerige (Property Tax), Child Labour Eradication Activities Information System (CLEAIS), KrishiMarataVahini, Ahara, Audit Monitoring System (AMS), RaitaMitra, CASCET—2003, E-Archive and many others to improve the telecommunication, human capital, infrastructure and C-Participation index.

4.7.1.3 Laggers

Nagaland stands first in the laggers list due to its poor performance in almost all E-Governance areas—online presence, telecom and E-Participation index. Broadband connections are weak, internet users are limited and interaction between government and citizens are low because the portal is not properly maintained.

Bihar scored lowest in Telecom and Human Capital Index which is attributed to a smaller number of broadband connections, mobile subscriptions, low literacy rate, limited gross enrollment ratio and less percentage of youth population in the State. Although Jankari, E-Gazette, Online Electricity Bill Payment, Online Enrollment in Electoral Roll and Information and Public Relations Department initiatives have been taken, implementation of the programs should be stressed upon.

Goa is the next participant in the list of laggers due to its low E-Participation Index, Online Availability and Performance Index, Telecom Index and Infrastructure Index. The State government needs to develop more awareness among the citizens about the available services and service delivery methods and means.

Meghalaya is at the fourth position attributed to its lowest online availability and presence index and a poor performance at telecom and E-Participation index. The State portal shows the availability of various E-Governance services such as Online VAT Application, online agriculture market price, Online Public Utility Forms and online constituency-wise electoral roll but the information is ambiguous.

Jharkhand is fifth in the laggers list due to a very low Telecom Index, a low Human Capital Index and E-Participation Index. The limited number of E-Transactions per service in the state and low literacy levels explain the poor performance in Human Capital and E-Participation index. The state needs to enhance the user interactivity levels of its E-Services and improve the literacy levels. The state government has implemented E-Governance programs E-Rahat Emergency Corner, Grievance Redressal—SamvadaurSamadhan, E-Nibandhan, Online Land Records, E-NagrikSeva, Common Service Centre and Gyanshila to improve the literacy and E-Participation levels in the state.

4.7.1.4 Potential Laggers

The north-eastern states have low infrastructure and telecommunications availability which explains the low infrastructure and telecom index in Assam. Apart from this, Human Capital and E-Participation index are also below average.

Rajasthan is second in the list of potential laggers. It has low telecom and E-Participation index and Human Capital and Infrastructure indices are below average. The state has launched E-Mitra, Raj stamps and grievance redressal facility through meeting with the Chief Minister but does not have even a single working SWAN despite being such a populous state. This is one of the major reasons it is not performing well despite of a large potential.

Manipur stands at the third position due to its low E-Participation Index, Online Availability and Performance Index and Telecom Index. The major problem faced by the state portal is the huge security issues which shows its gap in the online services and is the reason for its low participation index.

Chhattisgarh comes at the fourth position due to its poor performance in E-Participation, telecom and infrastructure. The infrastructure and telecommunication index need to be improved by improving the elements of citizen involvement and interaction.

The last state in this list is Jammu and Kashmir due to its low telecom, E-Participation and human capital index. The state experiences political stability, religious conflicts and demeaning acts like terrorism which makes it a potential lagger. The state has initiated E-Governance initiatives like Community Information Center—CIC, Online Employment Exchange

Information, Online Motor Vehicle Information and Information related to the Forms and Procedures but still is placed in the potential lagger list due to its weak administration and E-Governance.

4.7.2   Union Territories

4.7.2.1 Leaders and Potential Leaders

Lakshadweep tops the EGDII list of leaders among the Union Territories. The State has the highest Online Availability and Presence Index and highest Infrastructure Index due to the presence of active Common Service Centres and availability of E-Services. The Union Territory has also performed well on the human capital index but needs to improvise on the telecom index.

The next leader is Delhi due to highest Telecom and E-Participation index. But the UT has the lowest score in terms of online presence and infrastructure. High level of literacy, youth population and high gross enrollment ratio contributes to the success of the E-Governance initiatives of the UT despite low infrastructure.

Chandigarh has the potential to become a leader in E-Governance due to its significant performance in Online Availability and Presence Index and satisfactory performance in Human Capital Index and Infrastructure Index. However, it lags in terms of Telecom Index and E-Participation Index.

4.7.2.2 Laggers and Potential Laggers

Dadra and Nagar Haveli and Andaman and Nicobar Islands are the biggest laggers among the Union Territories. The UTs are the least performers in the Human Capital index, Infrastructure index and E-Participation index due to low literacy rate, governance and resources issues.

Andaman and Nicobar Islands have acclaimed second position in the laggers list due to its least performance in the Telecom Index, E-Participation Index, Human Capital Index and Infrastructure Index. The UT has shown satisfactory performance in Online Availability and Presence Index.

Puducherry is a potential lagger in the EGDII list of Union Territories. It has gained this status due to its below-average performance in E-Participation Index, Human Capital Index, Infrastructure Index and Telecom Index. It is because of the poor Broadband Wireline and fixed line phone connections. Although the UT has above average Literacy Rate and Gross Enrollment Ratio, it loses out on its Youth population declining its human capital index. Similarly, it has a high number of Operational and Planned Common Service Centres and number of E-Services, but due to low E-transactions per service, the Infrastructure Index and E-Participation Index is low. There is high availability of services and infrastructure in the Union Territory but the population is hesitant in using this infrastructure and these services which makes it a poor performer (Tables 4.11, 4.12 and 4.13).

A drastic change has been witnessed in the E-Governance potential of States and their E-Governance performance. Some states/UTs have underperformed as compared to their capability and resources such as

Table 4.11 Ranking of states

Ranking	States
Leaders	Himachal Pradesh
	Gujarat
	Uttarakhand
	Tamil Nadu
	Mizoram
	Lakshadweep
	Delhi
	Chandigarh
Potential Leaders	Kerala
	Andhra Pradesh
	Madhya Pradesh
	Maharashtra
	Karnataka
Laggers	Nagaland
	Bihar
	Goa
	Meghalaya
	Jharkhand
	Dadra and Nagar Haveli
	Andaman and Nicobar
Potential laggers	Assam
	Rajasthan
	Manipur
	Chhattisgarh
	Jammu and Kashmir
	Puducherry

Source: Gupta et al. (2016a, b, c, d, e, f, g, h)

Table 4.12 Ranking of states

Levels	Status	Environment	Readiness	Usage	E-Readiness
L1	Leaders (L1)	Karnataka, Chandigarh, Maharashtra, Andhra Pradesh	Karnataka, Chandigarh, Maharashtra, Tamil Nadu	Karnataka, Chandigarh, Delhi, Andaman and Nicobar	Karnataka, Chandigarh, Maharashtra, Tamil Nadu, Delhi, Andhra Pradesh
L2	Aspiring Leaders (L2)	West Bengal, Kerala, Haryana, Gujarat, Punjab, Tamil Nadu, Assam, Delhi, Madhya Pradesh, Uttar Pradesh	West Bengal, Kerala, Haryana, Punjab, Delhi, Bihar, Andhra Pradesh	West Bengal, Kerala, Haryana, Punjab, Delhi, Bihar, Andhra Pradesh, Gujarat, Tamil Nadu, Uttarakhand	West Bengal, Kerala, Haryana, Gujarat, Punjab
L3	Expectants (L3)	Goa, Orissa, Nagaland, Tripura, Puducherry	Gujarat, Andaman and Nicobar, Himachal Pradesh, Madhya Pradesh, Goa, Orissa	Maharashtra, Assam, Punjab, Haryana, Himachal Pradesh, Madhya Pradesh, Sikkim, Chhattisgarh, Rajasthan	Andaman and Nicobar, Madhya Pradesh, Goa, Orissa, Assam, Himachal Pradesh, Uttar Pradesh, Bihar
L4	Average achievers (L4)	Chhattisgarh, Himachal Pradesh, Sikkim, Rajasthan	Chhattisgarh, Uttarakhand, Jharkhand, Sikkim, Rajasthan, Uttar Pradesh	Orissa, Bihar, Goa, Uttar Pradesh, Meghalaya	Chhattisgarh, Uttarakhand, Jharkhand, Sikkim, Rajasthan
L5	Below-average achievers (L5)	Manipur, Bihar, Uttarakhand, Jharkhand, Andaman and Nicobar, Jammu and Kashmir	Puducherry, Jammu and Kashmir, Assam, Meghalaya, Tripura	Nagaland, Mizoram	Tripura, Nagaland, Puducherry, Meghalaya
L6	Least achievers (L6)	Meghalaya, Mizoram, Arunachal Pradesh, Dadra and Nagar Haveli, Daman & Diu, Lakshadweep, Arunachal Pradesh	Meghalaya, Mizoram, Arunachal Pradesh, Dadra and Nagar Haveli, Daman and Diu, Lakshadweep, Arunachal Pradesh	Tripura, Puducherry, Manipur, Dadra and Nagar Haveli, Daman and Diu, Lakshadweep, Arunachal Pradesh, Jammu and Kashmir	Manipur, Mizoram, Jammu and Kashmir, Dadra and Nagar Haveli, Daman and Diu, Lakshadweep, Arunachal Pradesh

Source: DIT & NCAER (2010); Gupta et al. (2016a, b, c, d, e, f, g, h)

Table 4.13 Comparison amongst States based on their E-Readiness (2008) and EGDII report (2015)

State and UTs	E-Readiness (2008)	EGDII report (2015)
Karnataka	1	13
Andhra Pradesh	6	10
Chandigarh	2	8
Haryana	9	18
Delhi	5	7
Maharashtra	3	12
Tamil Nadu	4	4
Uttar Pradesh	18	20
Punjab	11	14
Kerala	8	9
Rajasthan	14	25
Gujarat	10	2
West Bengal	7	21
Goa	14	31
Chhattisgarh	20	23
Himachal Pradesh	17	1
Madhya Pradesh	13	11
Jharkhand	22	29
Orissa	15	19
Mizoram	30	5
Puducherry	28	35
Sikkim	23	16
Uttarakhand	21	3
Meghalaya	28	30
Assam	16	26
Nagaland	26	33
Bihar	27	32
Andaman and Nicobar	20	27
Lakshadweep	33	6
Jammu and Kashmir	31	22
Tripura	25	15
Manipur	29	24
Daman and Diu	35	34
Arunachal Pradesh	31	17
Dadra and Nagar Haveli	34	28

Karnataka, Andhra Pradesh, Chandigarh, Haryana, Maharashtra, Rajasthan, West Bengal, Goa, Jharkhand, Puducherry, Assam, Nagaland and Andaman and Nicobar. With favorable environment, readiness and high ICT usage, these States were deemed to have E-Governance potential, but their current E-Governance status shows that the capacity has not been fully utilized. The states have failed to exploit their potential and lack in one or more following areas: E-Participation, web presence and information delivery, citizen interaction, infrastructure development, literacy, low E-Transaction per service, lack of Common Service Centres, weak administration and so forth.

On the other hand, many states have used their capabilities to enhance their E-Governance performance and performed better than what they were expected to be capable of. These states/UTs are Dadra and Nagar Haveli, Arunachal Pradesh, Manipur, Tripura, Jammu and Kashmir, Lakshadweep, Uttarakhand, Sikkim, Mizoram, Himachal Pradesh and Gujarat. This commendable performance is attributable to favorable infrastructural facilities, high literacy and gross enrollment rate, web presence, youth participation, better infrastructure and implementation of policies.

Tamil Nadu, Delhi, Punjab, Kerala, Madhya Pradesh, Meghalaya and Daman and Diu are states/UTs that have justified their performance by the E-Governance capabilities they possess. Their readiness status matches with their E-Governance performance.

Karnataka, Chandigarh, Maharashtra, Tamil Nadu, Delhi, Andhra Pradesh were found to be the leaders in E-Readiness (2008) with respect to favorable environment, technological readiness and ICT usage. E-Governance top performers in 2015 included Himachal Pradesh, Gujarat, Uttarakhand, Tamil Nadu, Mizoram, Lakshadweep, Delhi and Chandigarh. Delhi and Tamil Nadu are two states that have maintained E-Governance performance as per their readiness status.

The least performers identified in the E-Readiness report (2008) were Manipur, Mizoram, Jammu and Kashmir, Dadra and Nagar Haveli, Daman and Diu, Lakshadweep and Arunachal Pradesh while the E-Governance report suggests Nagaland, Bihar, Goa, Meghalaya, Jharkhand, Dadra and Nagar Haveli, Andaman and Nicobar are the lagging states. Dadra and Nagar Haveli being common in both lists. This suggests that Mizoram, Jammu and Kashmir, Daman and Diu, Lakshadweep and Arunachal Pradesh have performed well such that they do not appear in the laggers list of E-Governance while Nagaland, Bihar, Goa, Meghalaya, Jharkhand and Andaman and Nicobar have not utilized their available resources to improve their governance status.

CHAPTER 5

Changing Technological Trends for E-Governance

Saibal K. Pal

5.1   E-Governance Technology Trends

There has been successful implementation of National E-Governance Plan (NeGP) in the country. NeGP has been one of the most organized and successful program at both Central level and State level. Many of the Mission Mode Projects (MMPs) launched under NeGP have seen successful. These projects need advancement from various points of view like process re-engineering, need and feasibility analysis, appropriate project planning, and stakeholder's need analysis. The improvement is necessary as technologies are rapidly changing and the new ones like cloud, mobile, grid, and big data are emerging. These new technologies are now making an impact in various cases of E-Governance at the global level.

There is a need in further enhancement of NeGP in the form of "Digital India". E-Governance services are vital for almost all stakeholders, be it user (in rural or urban area) or institutional service provider at government level or private level. Mobile phone-based E-Governance services have been launched in recent past due to large-scale planning of E-Governance projects. M-Health, M-Education and M-Agriculture are some of the successful initiatives by the government using mobile-based E-Governance.

Various E-Governance projects have been launched and implemented by Indian government. In order to encourage future E-Governance projects, the Indian government is promoting technological advancements. The E-Governance projects involve technologies in pipeline such as Ubiquitous

S. K. Muttoo et al., *E-Governance in India*,
https://doi.org/10.1007/978-981-13-8852-1_5

computing, Free and Open Source Software, High Performance Computing, Blockchain Technology and Big Data Analytics. The future E-Governance projects such as Wireless Pollution Monitoring and Evaluation System, and Wireless Sensor Network for Real-Time Landslide Monitoring would use the above-mentioned technologies (Dutta and Devi 2016).

5.1.1   Ubiquitous Computing

Ubiquitous Computing involves mobile computing, distributed computing, sensor networks, context aware computing and location computing. The Department of Electronics and Information Technology (DEITY) is working on ubiquitous computing to generate intelligent products that connect to internet and data generated by it is easily available. CDAC (Hyderabad, Chennai and Bangalore), IIM Kolkata and Amrita University are the ubiquitous computing implementation agencies and academies in India (Dutta and Devi 2016).

5.1.2   Free and Open Source Software (FOSS)

Open source software allows users to modify or enhance the source code. It is available and freely licensed to use, change and share by a person. The National Resource Centre for Free & Open Source Software (NRCFOSS) is the official center to promote and encourage the development of FOSS community. One of the major initiatives of Free and Open Source Software is Bharat Operating System Solutions (BOSS). This system is Linux based that supports 18 Indian languages. Indian Navy, schools in Punjab and E-Governance programs of Chhattisgarh, Kerala, Puducherry, Haryana, Tripura, Tamil Nadu and Andaman and Nicobar Islands have deployed BOSS (Dutta and Devi 2016).

5.1.3   Lean Six Sigma

Lean Six Sigma process is proposed to be used in reform and re-engineering in order to achieve maximum efficiency. Lean Six Sigma process can be used to promote Justice, Transparency, Responsiveness, Equality and Accountability towards needs of the citizens. The use of Six Sigma in E-Governance can improve judicial processes. Lean Six Sigma in E-Governance can also be used in re-engineering Election processes as well as Legislation processes.

5.1.4   *Integrated Single-Window System*

Election Commission of India (ECI) has been working towards preparing an inclusive and integrated system for the election process in India by 2020. It will include various stakeholders of the election process. Election Management Bodies (EMB) all around the world are seriously considering the inclusion of ICT and other technologies in elections at various levels. Different seminars and workshops on technology and elections are considering and discussing the option of E-Voting along with various pros and cons about it. In several countries including India, discussions are underway with various stakeholders and reports are being prepared with different projects and strategies. Different technologies like cloud, mobile, cloud analytics and grid are being used for this process.

With the internet providing a way to integrate the voters and voting process, ECI and EMB are considering the option of online voting as well. Apart from being physically present at the voting center through technology, ECI is also considering the option of voting through the distant mode. Technology will help in enhancing the trust and transparency in the voting process (Press Information Bureau 2016).

5.1.5   *Blockchain Technology*

Blockchain Technology has received a widespread usage in the area of finance like that in managing crypto-currency. Usage of blockchain technology in the area of E-Governance has already started across many nations in the world (Martinovic et al. 2017; Jun 2018). The usage of blockchain technology has resulted in improved transparency, efficient working of the government and decreased complexity of the overall process. Various case studies have been developed related to financial transactions, land records registrations, birth registrations, tax collection, issuing of invoices, pension distribution and the likes. Government of India is also working on including the blockchain technology in their E-Governance framework for better and secured E-Governance services.

5.2   E-GOVERNANCE PLAN FOR NEAR FUTURE

In the past decade, various E-Governance initiatives have been launched. Vision of E-Governance based on global trends has been announced for the next five years for good governance. The vision includes following changes.

5.2.1   From Assisted Services to Mobile and Digitally Assisted Services

The increased use of smart phones and penetration of mobile phones and reduction in the cost of data has made it comfortable for the citizens to access government service from anywhere and reduced their travel time and effort to go to government offices or CSC for request of services or to file complaints. This can now be done with ease at homes with the help of mobile phones. Mobile phones are being used as a medium to provide all services, including those that are transactional in nature, on mobile by the government for process transformation and good governance (Fig. 5.1). For Transformational Governance, the combination of JAM (Jan-Dhan, Aadhaar and Mobile) needs to be focused upon and this combination is being implemented for better citizen engagement.

The use of mobile technologies in the national e-Government strategies at this stage is easier for the developing countries as they have just started progressing with respect to the services. Mobile technology helps in providing government services like grievance redressal, status of pending applications, filing tax returns, utility payments etc., and collecting feedback related to the government policies. (Nasscom 2016)

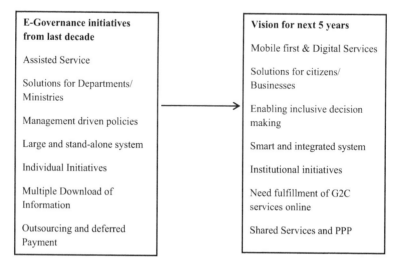

Fig. 5.1   E-Governance vision. (Source: Nasscom 2016)

5.2.2 From Solutions for Departments/Ministries to Solutions for Citizens/Businesses

The E-Governance programs generally focus on the digitization of services and processes of departments and ministries to enhance the citizen service quality. But the approach has some gaps such as citizens and businesses are not the center of concern in the approach and the approach limits integration.

The upcoming E-Governance programs are incorporating and integrating the interests of citizens and businesses. They are also converging the government departments digitally to provide better services to the citizens.

5.2.3 From Management-Driven Policy to Inclusive Decision Making

The E-Governance policies and initiatives are planned and implemented by the policy makers. However, there is a need for citizens to participate in policy and decision making for a better policy outcome. This can be done through both traditional and social media. Initially, the leaders/ managers were responsible for deciding policies, but the role has shifted to involve citizens in helping government to design policies. This change in decision making is reflected in the recent initiatives of the government in the form of MyGov, Mann ki Baat, and so on. Initiatives like RTI, CPGRAM have brought transparency and provided a perfect platform to enable this change. E-Participation should be made compulsory in decision-making policies (Nasscom 2016). Upcoming services will see a lot more inputs from the citizens for better policy making.

5.2.4 From Large and Stand-Alone Systems to Smart and Integrated Systems

From large and stand-alone systems in governance, the vision of E-Governance has shifted to smart and interconnected systems such as Cloud, Digital Media, and Mobile Internet. The technological advancement in India has significantly altered the way E-Governance is perceived and carried out and is moving towards Innovative solutions from Collaborative technologies. Adoption of smarter technologies and its alignment with Governance-IT is the next step.

5.2.5 From Individual Initiatives to Institutional Initiatives

Change of project leadership and individual orientation in E-Governance Projects often fails. Leadership sponsorship and individual initiatives needs to be separated. Leadership sponsorship strengthens the project while individual initiative makes the project brittle. There is a need to change individual initiatives to institutional initiatives as projects driven by individual initiatives cannot take changes smoothly. A change in project implementation teams results in change in requirements, understanding and scope of the project which hampers the implementation of the project.

5.2.6 From Multiple Download of Information to Need Fulfillment of G2C Services Online

Projection and tracking of services online is a complicated task. Rather than citizens and businesses coming to government offices or availing information/services online, the government should reach the needy citizens and provide support rather than people asking government for support (Nasscom 2016).

5.2.7 Outsourcing and Deferred Payment to Shared Services and PPP

To bring changes in the current model, there is a need to develop government-wide shared service centers to build capacity and increase efficiency. Partnership with private players will also scale up E-Governance.

India has been renowned as a power house of IT services all around the world and private sector has contributed to this growth. Now, the public sector of India needs to participate in the IT sector and aim Governance Transformation with the help of technology. For this, the government needs to push boundaries, set vision and take actions to incorporate IT in their strategies and decisions (Nasscom 2016).

5.3 PILLARS OF DIGITAL INDIA LEADING TO E-GOVERNANCE VISION

The vision of E-Governance program is threefold:

- Infrastructure as a utility to every citizen
- Governance and services on demand
- Digital empowerment of citizens

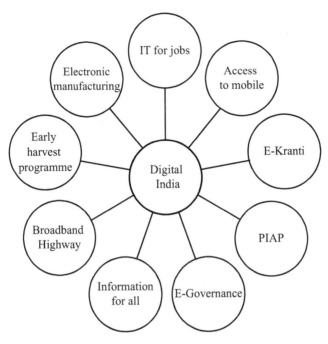

Fig. 5.2 E-Governance vision areas based on nine pillars of Digital India

The Digital India program under the E-Governance schemes in India has defined its nine pillars which forms the major vision areas to work upon (Fig. 5.2).

Initiatives introduced under the Digital India program include E-Sign, Skill India, PMJDY, JAM, E-Hospital, Wi-Fi hotspots, DBT, NOFN, Smart Cities and Digi Locker. These initiatives have been launched to bring a revolution in the governance and to ensure inclusive growth. Some of these initiatives have been successful in their initial phase and their future success is dependent on how policy makers deal with it along with the challenges they face in various stages. Some of these initiatives have been found to be encouraging enough.

5.3.1 Direct Benefit Transfer (DBT)

The DBT scheme has been launched to reduce malpractices and to ensure that the subsidies reach the needy people in fair manner. Under this scheme, consumers who are entitled to subsidies will receive subsidies directly into their bank accounts, while the subsidized goods will be available at market prices. This is done to reduce pilferage, adulteration and other malpractices by the sellers and to ensure that subsidies reach to the needy.

Progress Till July 2015 almost 140 million consumers benefitted from this scheme. The PAHAL initiative launched under this scheme was also successful. Over 1.62 million participated in this initiative and have voluntarily given up LPG subsidy under the "Give It Up" campaign of the government to serve the poor.

Technology Intervention for Success To avoid duplication of beneficiaries and to reduce leakages, it is suggested to link bank accounts with Aadhaar. This will allow efficient tracking and monitoring of benefits transfer.

5.3.2   JAM Trinity

The Jan-Dhan-Aadhaar-Mobile is another scheme launched for the benefits of citizens. This scheme proposed provision of three identification numbers to allow citizens to avail several government benefits. This initiative is meant to allow transfer of benefits in a leakage-proof and cashless manner.

Progress The scheme was launched in February 2015. But the implementation of scheme requires the use of Aadhaar. Since this proposal of involving Aadhaar is not quite feasible, the progress on this front is stuck.

Technology Intervention for Success Other identification modes, such as voter IDs that are more common than Aadhaar cards, can be used for authentication and to reduce dependence on Aadhaar cards. This will also help in efficient rollout of the scheme.

5.3.3   Smart Cities

Smart cities are cities with high-tech communication capabilities and infrastructure. The government aims to create 100 smart cities by 2022. In Surat, one of the cities selected for transformation as a smart city, a network of CCTV cameras has been set up to monitor crime. The central government has dedicated a sum of Rs. 480 billion for the development of smart cities.

Progress A list of 98 cities has been prepared to be developed as smart cities by the government. Offer for financial assistance has been provided by foreign agencies and development banks such as the World Bank, Asian Development Bank and KfW Development Bank (Germany) for the same purpose. Foreign nations are also offering their knowledge and expertise to share with the government. France has shown interest in providing assistance to the government for the development of Nagpur and Puducherry, the United States has vested interest in Ajmer, Vizag and Allahabad, while Singapore, Germany and Spain have exhibited interest in providing expertise for the implementation of the various initiatives. Public-private partnerships (PPPs) will have a great contribution in the implementation of the plan. Consulting firms are also participating in the execution by assisting in planning, strategizing and executing the plan. Based on the regional presence and expertise, consulting firms are delegated work in technology, strategy and infrastructure areas. Deloitte has been delegated with the planning and implementation in West Bengal, Bihar, Odisha and Andaman and Nicobar. Finally, the government has assigned capital for special purpose vehicles involving private firms and urban local bodies.

Technology Intervention for Success Technology can be used for development in areas of education, health, crime, energy, traffic and waste management and so on. Electronic delivery of government services and E-Platform for citizens will help resolving minor issues and encourage participation of citizens in governance.

5.3.4   *National Optical Fibre Network (NOFN)*

This initiative aims to connect the gram panchayats of the country. The program involves development of a high-speed digital highway to connect 250,000 gram panchayats using optical fiber. This is the world's largest rural broadband project.

Progress For the establishment and management of NOFN program, Bharat Broadband Network Limited has been set up. The program has been progressing slowly and for now around 1 percent of the gram panchayats have been covered under Bharat Broadband project and, therefore, the deadline for completion of the program is now 2019.

5.3.5   Wi-Fi Hotspots

This project aims to develop Wi-Fi-hotspots in the country to provide digital connectivity across India.

Progress Free public high-speed internet connection has been set up at several locations such as railways and metro stations. Cities like Mumbai and Delhi have Wi-Fi facilities.

Technology intervention for success High-speed high-intensity routers can be used to develop internet connection in public places. The electronics development fund can be used for the effective implementation of the program. India's initiative for becoming a zero net import country for electronics manufacturing under the Digital India campaign will help the implementation in a cost-effective manner.

5.3.6   Skill India Initiative

The aim of this program is to provide training to 400 million people belonging to different areas in India by 2022. Due to the lack of the skilled workforce, the success of the Digital India program is hampered. This initiative is, therefore, important for the success of the Digital India program.

Progress Fifty thousand youth in 100 job roles will be trained under the Pradhan Mantri Kaushal Vikas Yojana (PMKVY). SMS campaigns are used to train 400 million people. Recognition of Prior Learning is a government initiative to recognize and certify youth for their skills. The government also proposes to grant loans of INR 5000 to INR 15,000 to the youth for skill development.

Technology Intervention for Success The training centers must be equipped with high-speed Wi-Fi facilities and video facilities for enhancing the outreach and scalability of the project. Mobile applications can also be proved to be useful for skill development.

5.3.7   E-Hospital

Under the Digital India program, E-Hospital initiative has been launched and managed by Department of Electronics and Information Technology. This program involves an online registration system linking all the hospitals across the country. This program aids in the registrations and appointments based on Aadhaar. The initiative provides facilities like online registration, online diagnostic reports, fees payment, appointment and availability of blood online.

Progress The website—ors.gov.in allows tracking of data and registrations and also provides facilities like online diagnostic reports, fees payment, appointment and availability of blood online. The website is fully functional and is updated regularly.

Technology Intervention for Success Efficient database management can prove to be important to access facilities in other hospitals. Analytics can also be used for appointments and other queries which will allow hospitals to deal with resources in an efficient way.

Electronics Development Fund is an initiative that will allow venture capitalists to encourage research and manufacturing of electronics. This will help entrepreneurs to participate in the electronics manufacturing for the healthcare sector.

5.3.8 E-Sign Framework

This scheme will allow the users to digitally sign a document online with the help of its Aadhaar card.

Technology Intervention for Success E-Sign framework ensures the privacy and security of the documents. This feature will also allow rapid integration with government departments and procedures.

5.3.9 Infrastructure

5.3.9.1 Broadband Highways
Under the Digital India campaign, the government has allocated a budget of INR 5 billion for building high-speed broadband highways. This broadband connects all villages, departments, universities and R&D institutions. The broadband development contributes to the achievement of Millennium Development Goals (MDGs) with the help of fiber networks.

The National Optical Fibre Network (NOFN) launched by Universal Service Obligation Fund with the aim of providing broadband access to

250,000 gram panchayats of the country by 2016. BSNL, RAILTEL and PowerGrid Corporation are the PSUs responsible for laying 600,000 KM of fiber across the country.

Faster rollout of optical fiber network across the country would involve the participation of private players apart from the Public sector undertakings. Competition from private players is important to bring efficiency and to reduce the pricing of digital services on high speed. Quicker adoption of services related to bandwidth can be achieved through the innovative techniques and strategies for both operations, sales and marketing.

5.3.10　Digital Chip Maker

Digital Chip Maker Intel in association with the government unveiled multilingual training of digital skills and its application. There are multiple modules like financial awareness and inclusion, cleanliness, healthcare initiatives and digital literacy, are included in the scheme of Digital Chip Maker. Intel is working towards creating digital literates in 1000 panchayats with the help of the government.

5.3.11　E-Kranti

The project of E-Kranti aims to develop the provision of service delivery through electronic mode for the people of the country. There has been an allocation of around 5 billion Indian rupee for this project. E-Kranti includes sub-projects in the area of farming, planning, security, healthcare, education, law and justice through electronic mode.

5.3.12　Cloud Computing

Cloud Computing conceptually refers to the provision of various resources and services through a virtual set-up of the system, without the physical presence of the infrastructure at one place. It allows multiple users to access the services and infrastructure from multiple locations for multiple times. It also helps in dynamic allocation of the resources on demand over cloud platform.

The implementation of the Digital India requires the usage of cloud computing. For this there is work in progress towards formulation of policy related to its usage. Various strategic decisions are being brainstormed so that various services and infrastructure can be mapped to the cloud

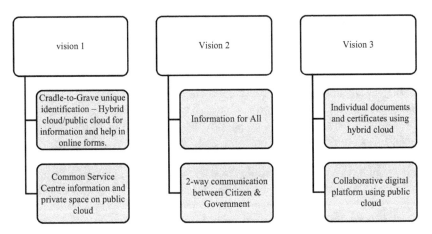

Fig. 5.3 Vision areas enabled by cloud computing

computing which can be provided to the different stakeholders. Architecture principle and data hosting policy needs to be worked upon for efficient delivery of services through the cloud model.

The data should be stored on variety of clouds. Like, the highly sensitive data should be hosted on private cloud, while the less relevant data and information could be hosted on a shared cloud service provider. Informatory services on the government platform could use open-access cloud services like those of Amazon web service or Google web services. Secured network can achieve multiple security checks through this way. Hybrid strategy could also be used for services which require integration, aggregation and customization (Fig. 5.3).

5.3.13   Service Enablement Support for 2G, 3G and 4G

Services would reach Indian citizens through network technology and spectrum. The bandwidth limitation existing across the country has an impact on the services build for public utilization. Network type (2G, 3G or 4G) also has an impact on the provision of services. An architecture design strategy must be developed to decide how services need to be deployed. Provision of Wi-Fi connectivity in government service centers using wireline broadband must be made to reduce the issues faced by the public associated with wireless networks.

5.3.14   Mobility: Web-API for Internal and External Consumption

A system based on 'System of Engagement' rather than on 'System of Records' needs to be developed. This information-centric strategy will enable collaboration and provide dependent services with the help of mobile apps, social media and analytics. This would require separation of data for internal and external consumption and linking of all the data with views and requirements of users.

Accurate information and data must be accessible in a secured medium and on all devices. For this, application should be built, and existing application and systems must be examined as information-centric resource. Data should be classified based on security policy for privacy, confidentiality, classified and open information and on the access to retrieve, create, update or remove (Fig. 5.4).

5.3.15   Security-Information Categorization

The digital architecture has security as an important component (Fig. 5.5). The development of digital infrastructure requires:

Organizational Planning—This involves structuring people, processes, data and information.

Security Principles—Policies and guidelines must be made in accordance with regulatory and security compliance.

Vision 1	• Mobile phone and bank account for financial inclusion
Vision 2	• Availability of Real-time service, digitally transformed services • Financial transactions going electronic and cashless
Vision 3	• Collaborative digital platforms
Pillar 5	• E-Kranti
Pillar 6	• Information for All

Fig. 5.4 Vision areas by mobility solutions

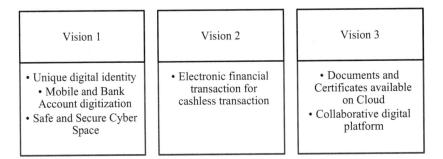

Fig. 5.5   Vision areas enabled by security

Architecture—Decisions regarding design patterns and standards.
Backend Application—Backend applications and existing assets must be assessed for their feasibility to digitize.

5.3.16   Analytics: Unique Digital Identity and Data Linkage with Security

A centralized secured database has been formed containing all details of the citizens regarding their address of residence, background check, utility consumption details, credit worth, and criminal records. UIDAI or PAN details are needed to connect with citizen details. The database with details of the citizens can then be made available to various parties and will be shared among services within public sectors.

Banks, financial institutions and other sectors are provided with details for analysis of their customers. The data available for individuals such as Aadhaar card, voter IDs, utility IDs, PAN, ration card and property tax can be used for background checks, credit worth, tax default, usage of electricity, gas, water and telephone, travel history, visa status, education qualifications, payment credibility, bank accounts, tangible and intangible assets and address. Analytics is highly useful for ensuring security compliance, confidentiality and privacy laws (Fig. 5.6).

5.3.17   Machine to Machine (M2M)

Machine to Machine (M2M) communications is a new way of information transmission that holds a huge opportunity since the mobile subscribers have grown to 900 million. M2M is quite useful for smart utility metering,

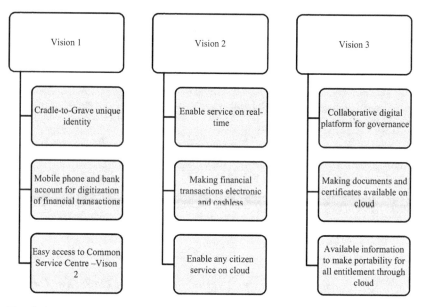

Fig. 5.6 Vision areas enabled by analytics

home automation, logistics, automotive, transport and supply chain and industry wireless automation. M2M plays an important role when machines/meters/logistics are linked to a central system. This connectivity ensures efficiency, accuracy and productivity of the mechanism. Connectivity in the machines within the sectors results in optimization of energy, utility and resources. M2M can help in resource optimization, revenue generation and to reduce revenue leakage. Communication Service Providers (CSP) provides proprietary radio-links, short-range radio signals or cellular-based connectivity that helps connecting machines, meters and resources and digitizing information. The M2M service allows integration of information and high-volume data transfer. With the help of M2M, data about individuals entering the government service centers can be accessed through their device and automatically processed in the background resulting in increased productivity and ease of doing business.

The data for GIS could be collected with the help of the user's identity and the device carried by him. The GIS can be helpful for connecting with machines or vehicles in transportation and even for mapping of the city, to develop decision support systems. There are a number of M2M technology-based applications as discussed below.

5.3.17.1   Connected Industrial Production

The working and health of any machine can be monitored remotely through connected industrial production. This will help in betterment of the working of machines as in case of breakdown of the rectification process can be applied at the earliest. In fact connected machines can send message to some other device in case of breakdown or non-working.

5.3.17.2   Connected Utility Meter

Various utility services like gas service, electricity service, telephones, and so on in households have meters installed to record the number of units consumed for respective services. To avoid any irregularities, breakdowns or tampering of these meters, centrally connected meters can be useful. Similarly, on vehicle tracking and toll collection can be done on national highways with the help of sensors and connected meters. Toll bills and traffic violation notices can be sent through the connected devices to the owner's E-Mail or phone.

5.3.17.3   M2M

M2M is an upcoming technology used for managing vehicles these days. Recently, electronic car REVA by Mahindra Company was built upon machine-to-machine technology for communication between various devices and its internal operations like central car locking system, emergency charging and air condition controlling. Message broadcasting is another application of M2M which can be executed at regional level, national level and even at the global level. Like for smart billboards displaying crucial information like weather updates or stock prices can change the data instantly with the help of M2M technology.

M2M can also be used in critical applications like space-crafts, nuclear installations, military operations, live surgery, and so on. Life-threatening events can be monitored and failure predictions can also be done with the help of M2M.

5.3.18   Social Media

The digital transformation is largely based upon social media. Social media was extensively used in the election campaign of 2014. People between the age of 34 and 60 years are the most active users. The usage of social media is improving with the increasing use of mobile phones and affordable data plans. Apart from entertainment, social media is also used to build awareness, help in social issues, bring people together, and enhance governance and employment.

Social media has helped the government in communicating with the citizens and vice-versa. Most of the government departments, agencies and officials are interacting with the citizens through the popular social media platforms like Facebook and Twitter. Citizens are using it to process their concerns and queries related to existing services and upcoming services as well. Even the ministers are using it to promote their departmental work and also to highlight their achievements. They are also directly communicating with a citizen which is giving confidence to the citizens towards the governance. Portal like 'MyGov' is also an initiative in this regard, which is making government pro-active in decision making and innovation. This is also helping in citizen inclusion for the decision-making process and reducing the communication gap between government and a citizen of the country.

5.4   Upcoming Projects

The department of Agriculture and Cooperation introduced AGRISNET as a mission mode project under National E-Governance Plan of Government of India. As a component of AGRISNET, "Strengthening / Promoting Agricultural Informatics & Communications" has been launched by Ministry of Agriculture. With the help of ICT, agriculture-related information or data can be extracted to improve upon its functioning and bring efficiency and transparency.

E-Biz is currently being implemented by the Department of Policy and Promotion (DIPP), Ministry of Commerce and Industry under Government of India. The project visualizes the transformation of businesses by providing them integrated, transparent and efficient electronic services online via a single window. This service is crucial for businesses, investors and industries to process the information on forms, procedures, license payments and compliances in the business cycle. The E-Biz project aims to transform the service delivery approach from department oriented to customer oriented. The first phase of the project involves offer of 24 services to various departments.

Various E-Governance projects are taking place that aim to flow authentication and application towards the next level. Considering security to be a necessity, these programs involves multiple options for security like credential login processes, Security question-based authentication, Security pin number, Biometric authentication and digital signature-based processes. Digital Signature has been recognized as an important part of authentication and, therefore, has made its place in the E-Governance projects. A Digital Signature Management Cell (DSMC) needs to be established to facilitate the functioning of Digital Signatures across the State departments. The DSMC may have the functionality related to Digital Signature, Authentication, submission of application, Procurement process and handover of Digital signature to the applicant.

IPv4 is a 25-year-old internet protocol with the capacity of 4.3 billion IP addresses. The current version of internet protocol IPv4 is getting outdated with the report growth of internet in coming years. With the advancement in the internet, Internet Protocol version 6 (IPv6) was developed by the Internet engineering Task Force (IETF), which is an advancement on the address capacities of IPv4 which makes available an infinite pool of IP addresses.

The state requires an immediate action to be taken to identify IPv6 compliant and non-compliant equipment's and software for the implementation. A procurement plan for IPv6 compliant hardware and software also needs to be made by the government. A state-wide action plan also needs to be made to prepare all networking hardware, software and all website to adopt IPv6. There is high scope for the states to participate in the Internet Engineering Task Force (IETF) standardization effort for the infrastructure required for emerging 6TiSCH.

The E-Court Mission Mode Project launched under National E-Governance Plan aims to incorporate technological perspective in the district/subordinate courts of the country. The core objective of the project is the ICT enablement of the courts and provision of designated services to litigants, lawyers and the judiciary with the help of ICT. The basic infrastructure for ICT enablement has already been developed, such as installation of computer hardware, Local Area Network (LAN), internet connectivity and connection of application software at complex of each court. The courts were also provided with facilities like laptops, laser printers, broadband connectivity, and judicial officers were also provided ICT training.

For further technological penetration, Universal computerization of all the courts, use of cloud computing, digitization of case records of last 20 years and enhanced availability of E-Services to lawyers and litigants through E-Filing, E-Payment gateways and mobile applications have been adopted.

The Phase II of the ICT enablement of E-Courts project has been approved and a budget of INR 1670 Crores for the project has been decided. The project comes under the purview of Digital India program of the Government of India. The Phase II of the project involves computerization of courts plus introduction of the automation of work-flow management that would enhance the control over the management of cases. Phase II involves a step further towards ICT enablement of the courts which will include installation of touch screen-based Kiosks with printers in each court complex, use of mobiles to fetch information, implementing change management and process re-engineering in courts, video conferencing facility at all court complexes and corresponding jails, use of E-Filing, E-Payment and mobile applications. These services would be provided through the judicial service centers. Hand-held process service devices would also benefit the judiciary and process servers to enhance transparency and time-bound delivery of court notices and summons. Phase II also encourages the use of Digital Signature Certificates (DSCs) for court officials in issuance of E-Documents to lawyers and litigants. Digitization would also benefit in areas like court management system document management, Judicial Knowledge Management and Learning Tools Management. And for the courts to contribute to the environment, the use of solar energy has also been proposed at complexes of courts.

Since the project comes under the Digital India campaign of Government of India, the objective of the project falls in line with the objective of Digital India Program of the Government of India emphasizing on citizen-centric services providing governance and services on demand to every citizen and digitally empowering them. Data analytics will be used by the government to process ideas generated by various E-Governance portals and to use it for better governance.

5.5   Focus Areas

5.5.1   Digitally Green Agriculture

The Indian agricultural sector contributed around 15 percent to the GDP in the year 2013–2014. More than 50 percent of the population is employed in the agriculture and allied sectors. The rapidly increasing population needs to be fed and, therefore, agriculture should produce more. To accelerate the growth of agricultural sector, ICT tool should be utilized to boost the economic growth of the country. The lack of knowledge about new technologies and initiatives by the government makes the productivity of farmers low. Social media can help in removing middlemen and connect buyers and farmers. For example, as per a Deloitte study,

Facebook has a global economic impact. Facebook enabled $4 billion economic impact and created 335,000 jobs in India in the year 2014. Facebook groups have been formed where farmers buy and sell to meet the demand-supply gap. ICT tools like M2M can be used to improve productivity by appropriate use of pesticides, fertilizers, and other farm resources based on soil detection, weather conditions and so on. Geographic Information System (GIS), Geographic Positioning System (GPS), monitors and controllers for agricultural equipment can allow farmers to direct equipment movements using electronic guidance. Precision farming is another ICT tool which helps in various agricultural activities such as yield monitoring, yield mapping, variable rate fertilizer, and variable spraying. Soil Health Card software and web-based software for nutrient management are used under the E-Governance program for eight states (Deloitte 2016).

5.5.2   Education

Education enhances knowledge and skills and helps in socio-economic transformation of the country. The Indian education market was worth $92.98 billion in 2014. From FY05 to FY12, the sector grew at a CAGR of 16.5 percent. Social media and mobile phones have provided an efficient platform to provide education in an efficient way. The industry is flourished with innovative ideas to educate the masses in an economical way. These innovative ideas include Fisher Friend, Pragati, English Seekho, iPerform and so on. Mobile education is growing fast, and it is estimated that it will be a $70 billion market in 2020.

The government has assigned a budget of INR 1 billion for developing virtual classrooms and online courses. Virtual classes and online courses need strong data connectivity and improved IT infrastructure. The Massive Online Open Course (MOOCs) are easily accessible and have no restriction on class size which makes them easy to adopt (Deloitte 2016).

5.5.3   Healthcare

The Indian healthcare sector is growing at the rate of 15 percent CAGR and the industry is expected to grow from $78.6 billion in 2012 to $158.2 billion in next five to eight years. Although the growth rate is promising, the industry still has many faults such as a lack of proper infrastructure, workforce, unequal access to healthcare facilities and high healthcare costs.

These issues in the healthcare sector can possibly be resolved with the help of digital transformation of the healthcare sector. Various digital facilities need to be introduced that can change the face of the industry such as hospital information system (HIS), electronic health record (EHR) and picture archival and communications systems (PACS). Some of these initiatives have been introduced by the Indian government under the E-Governance plan but proper implementation is required to allow change and improvement in the sector.

Telemedicine is the new trend in the industry which is growing with CAGR of 20 percent and will serve as a solution for rural and remote people. Telemedicine is the solution for higher costs of specialized doctors that visit rural areas from urban areas. Telemedicine allows remote communication between physicians and patients. Very large population of India has no access to affordable medical treatment today. Affordable primary medical and diagnostic care will be accessible with E-Visits and necessary infrastructure.

Cloud technology can be used by the doctor to store and access data from anywhere, anytime and to deliver a real-time solution to patients' problem. The medical database of the required check-up data can be viewed directly by the patient through internet connectivity.

The M2M healthcare device market in India is growing with a CAGR of 33.81 percent from 2011 to 2016. It is a highly valued sector aiding in many medical cases. For instance, in the condition of heart arrhythmia, M2M-based heart-monitoring device records the daily condition. The recorded information can be transferred with the help of a mobile network or internet so that doctors or care-takers can review the person's cardiac condition and provide support accordingly. Many such M2M-based devices are already being used worldwide. Its usage must be enhanced in India as well. Real-time location and Global Positioning System (GPS) are other ICT tools that can be used in medical industry. These tools help to track the ambulance and suggest the shortest route to hospital. Dementia and Alzheimer patients can also be tracked by GPS (Deloitte 2016).

5.6   Security Concerns

Under the E-Governance plan, cyber security concerns play a major role. Cyber security requirements need to be changed with the change in threat environment. Regular updates are required in threat landscape in order to avoid emerging attacks. Sharing of information regarding emerging threats and vulnerabilities can help prevention of cyber-attacks. This can be done through collaboration of various agencies. A holistic approach is required to secure Indian Cyber Space. The cyber security strategies of the XI Five Year Plan will be implemented and improved, while new initiatives would be introduced to tackle with emerging threats and changing technology picture (Ministry of Electronics and Information Technology 2016).

During the XII Five Year Plan, some new cyber security strategies have been planned and projected:

(a) The policies need to be altered according to the changing threat landscape, complex cyber space and convenience of obtaining resources in the area of cyber security.

(b) Public-Private Partnership is a solution for security issues and for prevention of security issues.

(c) Security infrastructure, skills of technical employees and enhancement of awareness is required by country's cyber space to prevent cyber-attacks, and to minimize the national vulnerabilities to cyber-attacks.

(d) Collaboration and interaction between important stakeholders such as Government sector organizations, sectoral CERTs, International CERTs, product and security vendors, NGOs security and law enforcement agencies, service providers including ISPs, academia, and media, and cyber-user community is required to prevent cyber-attacks.

(e) Mock drill must be conducted to prevent cyber crimes and to assess the preparedness of sector organizations about whether they would be able to resist cyber-attacks and improve the security posture.

(f) Research and technology demonstration, proof of concept and test bed projects in areas of cyber security linked with recognized R&D institutions must be supported by the government.

5.6.1   Focus Areas

Six areas of cyber security will be focused by the government during the XII plan period:

(a) Legal framework,
(b) Security policy, compliance and assurance,

(c) Research and technology in cyber security,
(d) Concentration on early warning and response under security incident,
(e) Security awareness, enhancement of skills, and
(f) Public-Private Partnership.

Legal Framework

In order to enhance the legal framework of cyber security concerns, research projects related to cyber laws and related aspects like, E-Commerce, encryption, IPR issues, and privacy need to be conducted. Such research projects would provide insights about the cyber laws and issues related to the cyber security which would help in creating a better legal framework. Moreover, a database of legal cases related to cyber fraud decided in India must be created. A policy and procedure need to be devised for extraction of authentic data stored in repositories and hosted by Indian companies on servers abroad for access for lawful purpose. An encryption/decryption framework is also essential considering the concerns of both industry and Law Enforcement Agencies.

Security Policy, Compliance and Assurance

The compliance and assurance of security policy requires following below-mentioned measures:

(a) Cyber security studies and surveys needed to be conducted annually;
(b) Crisis management plan needs to be developed and implemented;
(c) Security audit and assessment needs to be done and certification infrastructure has to be enhanced. This involves enhancement of third-party certification, empanelment and ratings of auditors, cyber security drills, technical security testing and self-certification;
(d) A national cyber security index needs to be generated which would lead to national risk management framework; and
(e) IT product technical security assurance mechanism needs to be enhanced which would improve Common Criteria security test/evaluation and Crypto Module Validation Program.

Cyber Security Research and Development
Research and development will be carried out with basic focus on setting up test beds, development and demonstration of technology, transition and commercialization of technology. Public-private partnership needs to be used for joint R&D programs. A joint effort of industry and universities is important for the implementation of the activities. The major steps proposed in this respect are as follows.

(a) Taking efforts to set up center for excellence in Cryptography, Mobile Security, Malware Research, and Cyber Forensics
(b) Technology transfer will be promoted and prototype to production of products will be facilitated
(c) Cryptanalysis, algorithm design/development/hardware realization will be implemented
(d) Programs for attack detection, protection, recovery and prevention will be installed
(e) Security solutions for cloud environment
(f) Mobile security solutions
(g) Security requirements in SCADA systems will be fulfilled
(h) Assurance framework of Cyber security for the government sector will be introduced

Security Incident: Early Warning and Response
To rapidly respond to the security incidents and to facilitate information exchange required for cyber security, National Cyber Alert System will be strengthened. Development of infrastructure and use of secure computer and communication networks are the priority actions required to be taken by the government to strengthen the IT sector.

The major actions include:

(a) Establishment of Threat, Vulnerability and Malware Research Centre has been proposed
(b) Improvement of CERT-In Operations
(c) Development of sensor/honeypot networks at key ICT installations
(d) A knowledge repository at the central level

(e) Development of a response mechanism at national gateways
(f) In order to promote authority and accountability related to cyber security defense measures, Security-Information Sharing and Analysis Centres (ISACs) Cyber Security Operational Centre (CSOC) will be set up to enhance coordination with regional level Cyber Security Help Desks
(g) Development of Botnet Cleaning Centres in the government, infrastructure and public sector organizations.

Security Awareness, Skill Development and Training
There is a need to develop capacity, skills and training mechanism to build a strong cyber security workforce. Human resources are needed to cope up with the security challenges arising at government as well as private sector. Skill upgradation and retraining of existing employees must also be promoted.

The proposed measures include:

(a) Mass awareness campaign to create awareness related to cyber security among citizens is to be launched.
(b) Electronic media is to be used to create awareness.
(c) Cyber Security Training Labs/facilities will be developed across the country.
(d) Development of examination, accreditation and certification infrastructure.
(e) Cyber Security Concept Labs, Cyber Security Auditing of Assurance Labs, Digital Cyber Forensic Training Labs and SCADA/embedded security labs have been proposed to be developed.

Collaboration
Global cooperation is required to enhance cyber security. Shared understanding and interaction among agencies are the keys to mitigate cyber-attacks. CERT, law enforcement agencies and global agencies must coordinate to mitigate cyber threats. A well-established cyber security collaborative framework is required to be developed with the collaboration of government, private sector, partners, academicians, and national and international agencies. The Department of Information Technology should take charge of the collaboration for cyber security facets.

(a) Overseas CERTs and industry must be coordinated to promote cyber security
(b) Preemptive involvement at UN and Asia-Pacific level
(c) Awareness mechanism will be introduced within the country
(d) Coordination with law enforcement agencies and judiciary
(e) Development of a tiered structure for information sharing
(f) Collection of cyber security policy discussion and decisions

So, the major targets for cyber security focused in the XII Five Year Plan include identification of gaps in the existing policy and addressing them, development of national cyber security index for risk management, development of centers of excellence for advanced Cyber Security R&D, net traffic analysis, formal security education and awareness programs and promotion of collaboration between national and international agencies on cyber crimes and security (Ministry of Electronics and Information Technology 2016).

CHAPTER 6

Case Study Analysis

Rajan Gupta

6.1   Introduction

Nowadays, technology plays an indispensable role in an individual's life. Everything seems to be just a click away, due to the increased usage of different devices equipped with internet. Lately, India has observed a technological paradigm shift due to the widespread inclination towards the different electronic modes coupled with the innumerable government initiatives. Further, adaptation of the technological changes has marked an exponential growth in every sphere of the Indian industry, as it has been realized as one of the key elements for economic growth. The different technology-aided products and services are being used to cater domains such as *education, BFSI, healthcare, ITeS, manufacturing, defense, medical.*

The use of technology has showcased a humongous growth in government sector also. Few instances to support this include online *LPG bookings, payment of electricity and water bills, availability of birth and death certificates, passports,* and *railway ticket booking.* Furthermore, the advent of Aadhaar-linked services to facilitate the smooth functioning of the system and reduce the idle time of the people/customers has led to an increase in the convenience levels. The rise in the use of biometric systems and various computer-based learning methods in the organization are few more instances wherein the use of technology has evolved with time.

All this has led to the formulation and implementation of the E-Governance activities. E-Governance in India has been consistently evolving, which entails the computerization of the government departments

© The Author(s) 2019
S. K. Muttoo et al., *E-Governance in India*,
https://doi.org/10.1007/978-981-13-8852-1_6

and the initiatives that encapsulate the crucial aspects such as *citizen centricity, service orientation* and *transparency*.

Furthermore, good governance revolves around the parameters of *participation, data transparency, fairness* and *accountability*. With the rising advances in the communication technologies and internet-driven applications, aiming at the complete transformation of the people and its government in a new way is contributing to the accomplishment of *good governance*.

E-Governance primarily comprises of the implementation fields such as E-Administration, E-Services and E-Democracy. *E-Administration* focuses on the improvement of the different government processes and the public sector with the help of the various digitally executed information processes. *E-Services* entail the delivery of the public services to its citizens and *E-Democracy* emphasizes on the greater citizen participation and involvement with the help of the different ICT-enabled decision-making processes (UNESCO 2017).

6.1.1   Digital Profile of India

6.1.1.1   Investment in the IT Sector

India is considered as one of the most premier destinations for global IT and IT-enabled outsourcing services. The IT sector has played a vital role in the overall development of the Indian economy. The key drivers for the growth of the IT sector include cost efficiency, diversification across new verticals, inclusion of different pricing models, and optimum utilization rates (Invest India 2017). In the Union Budget declared for the FY 2017–2018, the Government of India has allocated an amount of Rs.10,000 crore for the Bharat Net Project aiming at the provision of high-speed internet connectivity to more than 150,000 Gram Panchayats (LiveMint 2017).

Prime Minister Narendra Modi launched the Bharat Interface for Money (BHIM) application, an Aadhaar-based mobile payment application which supports payment via credit or debit cards, increasing the numbers of paperless transactions. For the period 2014–2016, investments made in computer hardware and software has been 500 percent. Furthermore, exports made in IT sector have showcased an upsurge from USD 97 billion (FY 2014–2015) to USD 108 billion (2015–2016). The IT sector has witnessed the creation of over 2 lakh jobs in the year 2015–2016. The overall investment growth in IT sector is estimated to be 67 percent.

6.1.1.2 Increase in the Number of Internet and Smart Phone Users

The numbers of internet and smart phone users have exhibited an exponential growth contributing towards the creation of robust technological framework. In 2014, India accounted for about 24.3 crore internet users which have marked a huge growth for the year 2016 and accounted for about 50 crore.

In 2014, the number of smart phone users registered was about 22 crores which have increased and reached up to 37 crores for the year 2016. This simply directs towards an increased inclination of the users towards the various internet-equipped devices and development of a healthy technology ecosystem.

6.1.1.3 Rise in the Registered Aadhaar Card Users

The year 2014 registered about 73 crores Aadhaar card users which have shown a manifold increase and reached up to 109.5 crores. The Unique Identification Authority of India manages the Aadhaar numbers and Aadhaar identification cards.

For instance, a scheme called *Zero Balance Selfie Account* introduced by Federal Bank requires few easy steps to facilitate the opening of an account. It is a mobile-based facility that allows the users to open savings account through the bank's own platform called *Fedbook* which enables the users to open their accounts by scanning their Aadhaar cards or PAN cards and easy updating of passbooks. Currently, it is available on Android or Apple smart phones (Federal Bank 2017).

6.1.1.4 UN E-Governance Index Reveals India Climbs 11 Positions

In comparison to India's position of 118 in 2014 and an E-Government Development Index (EGDI) of 0.3834, India has recorded a rank of 107 and an E-Government Development Index (EGDI) of 0.4638 for the year 2016. This clearly indicates the rise in the number of the one-stop online platforms by the country (UN.org 2017).

6.1.2   Digital India Initiative: Power to Empower

As technology has made its presence felt in different business sectors, and to suffice the growing needs, numerous initiatives have been launched. The creation of a digital infrastructure, availability of the digital services and increase in digital literacy among the citizens has led to the Digital

India initiative primarily focusing on the power to empower. *The initiative was launched by our honorable Prime Minister Mr. Narendra Modi on 1 July 2015 as a step to digitally empower the nation.*

6.1.2.1 Vision of Digital India Initiative Revolves Around Three Aspects Namely

6.1.2.1.1 Digital Infrastructure as a Utility to Every Citizen
The digital infrastructure covers the parameters such as the access to the different internet-enabled services, knowledge pertaining to the various financial services available, easy access to the various common service centres and a safe and secured digital framework.

6.1.2.1.2 Governance and Services on Demand
The seamless unification of the digital services across the different departments and jurisdictions, availability of mobile-integrated services, such that the user authentication could be maintained along with the transparency, coupled with the timely information of the data transactions taking place.

6.1.2.1.3 Digital Empowerment of the Citizens
By digital empowerment one means the availability of the universal digital literacy along with the accessibility to the different digital resources such as Automated Teller Machines (ATMs), kiosks, POS (Point of Sale), POP (Point of Promotion), and so on. Further, encouraging the citizens to maintain their E-KYC (Electronic-Know Your Customer) so that it would be easier to link different bank accounts at a common place. Since the inception of E-KYC, the numbers increased from 0.03 crores in March 2014 to 40.63 crores in December 2016. All the activities are aligned towards the protection of the individual identity, prevention of data theft, money laundering and financial fraud (MEITY 2017).

6.1.2.2 Nine Pillars for the Digital India Initiative

6.1.2.2.1 Broadband Highways
This aspect focuses on three levels, viz., rural, urban and national. It aims at the provision of high-speed connectivity and cloud services and solutions across the various levels. Further, the integration of the different network and cloud services across the different government departments up to Panchayat level will aid in the overall augmentation of the network infrastructure across the country.

6.1.2.2.2   Universal Access to Phones

India has about 55,619 villages in the country that do not support mobile coverage. Large parts of the north-eastern states lack the adequate coverage and network resources. Thus, Digital India initiatives aim at reaching such underprivileged regions and provide them the desired level of network resources to develop the present prevailing conditions. Further, the Department of Telecommunications will be the nodal department and project cost will be around INR 16,000 crores during 2014–2018.

6.1.2.2.3   Public Internet Access Program

The availability of the Common Services Centres (CSCs) and Post Offices, basically act as multi-service centers to facilitate ease in performing the different tasks.

6.1.2.2.4   E-Governance

This aspect aims at reforming the various Governance activities with the aid of technology.

6.1.2.2.5   E-Kranti

To ensure the government-wide transformation by the delivery of the various government services electronically to the citizens with the help of multiple platforms.

6.1.2.2.6   Information for All

This refers to the delivery of adequate levels of information to all, such that timely decisions could be made.

6.1.2.2.7   Electronics Manufacturing

The National Policy of Electronics (NPE) launched in 2012 emphasizes on the creation of a conducive environment to attract the domestic and global organizations to make investments in the growing Electronics Systems Design and Manufacturing (ESDM) in India. The targets are to achieve net zero imports of the different electronic devices.

6.1.2.2.8   IT for Jobs

This pillar primarily focuses on the provision of adequate training to the youth to nurture their skills and make them job-ready for the various employment opportunities in the IT and ITeS sectors.

6.1.2.2.9   Early Harvest Programs
It refers to the programs that could be implemented within a short span of time like the IT platform for messages, government greetings to be E-Greetings, biometric attendance, National Portal for lost and found children, public WiFi spots, and so on (Digitalindia.gov.in 2017).

6.1.2.3   Recent Activities to Support the Digital India Initiative
8 March 2017: Hewlett Packard launched a Centre of Excellence (CoE) in support of the Digital India initiative

HP Inc. has introduced this platform to provide digital solutions to the various real-life problems and thereby enable the digital transformation of the country. The CoE was inaugurated by P. P Chaudhary, the Union Minister of State for Ministry of Electronics and Information Technology and Ministry of Law and Justice which emphasizes on the comprehensive digitization of the nation. The 4000 square feet facility has been developed in collaboration with 25 system integrators comprising of various other IT software companies. Further, it aims to offer solutions for numerous sectors such as Banking and Financial institutions (BFSI), education, healthcare and manufacturing.

15 March 2017: The Government planned to introduce the Start-up India hub most likely in April

The virtual Start-up hub introduced by the Department of Industrial Policy and Promotion (DIPP) being developed by Invest India was launched by the first week of April. The portal was supposed to integrate and replace the Skill India website.

17 March 2017: BHIM App crossed 18 million downloads

BHIM or Bharat Interface for Money is a mobile-based digital solution application. It is aimed at increasing the number of cashless transactions. Further, the application provides its users safe and secured transactions and thereby prevents the chances of data theft.

29 December 2015: Government launched 22 new schemes under Digital India Program

The schemes introduced encompass areas such as digital infrastructure, digital empowerment on-demand government services and promotion of industry. The services launched included electronic payments which supported the various government services, a Geographical Information System (GIS), a Request for Proposal (RFP) to allow the selection of the private cloud service providers for the government departments, an online laboratory specifically designed for the CBSE students to perform the experiments online, the incubation of ten projects in the area of chip-to-system design and the development of a native operating system that will support text-to-speech technology for nine regional languages.

6.2   METHODOLOGY

In the following chapter, we have discussed 12 case studies revolving around the aspect of E-Governance in India. The different case studies discussed in this chapter are *National Portal of India, Aadhaar, Mobile Seva, MyGov.in, E-Transaction Aggregation and Analysis Layer (E-Taal), Mann Ki Baat, Jeevan Pramaan, Pay Gov, Pradhan Mantri Jan Dhan Yojana, Digital Locker, National Center for Geo-Informatics, and National Scholarships Portal.*

The above-mentioned initiatives under E-Governance have gained huge likeability and popularity since their inception, owing to the numerous benefits they offer such as user-friendly platforms, facilitate decision making and help to complete the tasks timely. Moreover, platforms like Pradhan Mantri Jan-Dhan Yojana (PMJDY) and *PayGov* has helped in having paperless transactions, which reduced the hustle and aid in leveraging the overall efficiency.

The case studies discussed in this chapter have been arranged based on their launch dates and have been divided into sub-sections which entail an overview, performance evaluators, and a case summary providing the crux of the case study and future aspects related to it.

The framework for the case study analysis is shown in Fig. 6.1. The analysis consists of three major portions, which includes the overview, performance evaluators and the case summary. The first sub-section, that is, overview of the case, focuses on the introduction of the case by giving its background, various objectives related to it and different vision/mission statements if associated with that scheme. Performance evaluators includes the statistics related to the scheme like number of users engaged or number of services delivered, depending on the type of the scheme. It also covers up the implementation coverage and few subjective parameters like the features and benefits derived out of the given scheme based on its good performance. The last sub-section presents the case summary, where the conclusion based on statistics is presented, along with the impact of the mentioned scheme. The case study analysis varies for different schemes as features and offerings are completely different for various case studies under E-Governance in India.

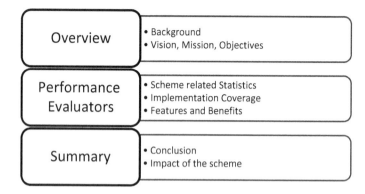

Fig. 6.1   Overall structure of the case studies being analyzed in this chapter

6.3   CASE STUDIES

6.3.1   *National Portal of India*

6.3.1.1   *Overview*

Information and Communication Technology (ICT)-enabled products and services are being adopted globally by individuals, companies and governments. This adoption of ICT is meant to reinvent the routine functioning of departments, to alter processes and service dissemination methods. There are 7000 Indian government websites and portals existing in the internet space as of today and this number is constantly increasing to fulfill government motives and objectives.

The purpose of launching a national portal is to bring transparency in the system, provide access to information and services to all citizens. National Portal of India (NPI), that is, www.india.gov.in was introduced under a Mission Mode Project of National E-Governance Plan (NeGP) to provide single window for online information and services to Indian citizens.

The National Portal of India serves the need of information to citizens of India and serves as a host to over 1900 services from various states/ministries/departments. The 'How Do I' section of this portal covers all the 1900 services. NPI is a central repository to store important government documents, schemes, forms and policies. The portal solicits feedbacks and suggestions, from users to make improvements in the portal.

Apart from providing transparency to the system and fulfilling objectives of E-Governance, http://india.gov.in has been directed to initiate a few ICT programs such as Sixth Central Pay Commission Public Consultation Process, the Central Information Commission Online (CIC Online), NGO Participation System (NGO-PS) and Public Partition on National Testing Scheme (NTS).

National Portal of India works towards strengthening the governance of the country by delivering government information and services to the citizens, businesses and stakeholders in India and abroad. The portal provides a unified interface and initiates all E-Government programs. The aim is to build a healthy relationship between the government and the citizens by aligning efforts of the various government authorities.

6.3.1.2 *Performance Evaluators*

6.3.1.2.1   Benefits Delivered (G2C, G2B, G2G, G2E)
The aim of this project was to develop a single-window access to information and services for citizens of India and to enhance interface between Government and Citizen (G2C), Government and Business (G2B), Government and Employee (G2E), and among Government Departments. The portal acts as vital source of practicing and enhancing good governance.

The Portal, 'http://india.gov.in' has a nationwide reach in the sense that it helps in transferring information to a large audience, even remote areas, to districts and Panchayats. The beneficiaries of information include following.

(i) All Indian citizens irrespective of age, gender or any demographic background
(ii) Government departments, ministries and associated offices
(iii) State Governments, Union Territories, Districts and Panchayats
(iv) Legislative and Judicial Institutions
(v) Public and private sector organizations
(vi) National and International Media Agencies
(vii) Non-Resident Indians (NRIs)
(viii) General Public all over the world

6.3.1.2.2   Stakeholders and Benefactors

G2C

The portal fulfills the primary objective of E-Governance by facilitating interaction between government and citizens. It allows electronic delivery and exchange of information. Services made available by the portal include obtaining certificates, renewing licenses, paying taxes/bills and accessing benefit schemes in a convenient manner. The portal also allows citizen participation in policy formulation.

G2B

The portal facilitates interaction between government and businesses regarding licenses, taxation and policies made for respective sectors and industries. The interaction with the businesses has led to enriched procurement of goods and services by the government. The terms for entering and exiting a business have been provided in an understandable manner. The Online Government Tender Information System has also been integrated with the portal.

G2G

NPI allows interaction and sharing of data within government departments. There is intra- and inter-exchange through NPI.

G2E

Government-to-Employee interaction is also facilitated by NPI to exchange information related to employment opportunities, rules and regulations, compliances and benefits of the job. A special corner has been dedicated namely 'Employee Corner' in NPI to fulfill information needs of prospective and currently working government employees. This column provides up-to-date information of Central Pay Commission.

6.3.1.2.3   Services Introduced

Single-Window Service

Various government bodies provide services that are spread over several websites and, therefore, many of them remain unnoticed. The portal ensures that the services are easily accessible to all stakeholders. The large amount of information placed on the portal is organized with the option of state-of-the-art search.

Comprehensive Content
The portal contains comprehensive information about all aspects of central, state or local bodies.

Bilingual Content
The content on the portal is of bilingual nature, that is, available in both Hindi and English. The sections in the portal are available in Hindi and they will soon be made available in regional languages also. The purpose of making it available in many languages is to widen its reach.

6.3.1.2.4   Implementation Coverage
The portal has experienced development in the service delivery over the years. It has the capability to attract visitors by making updated information and services available to them.

Enhanced Services and Interactivity
To fulfill the very objective of NPI, over 1900 services have been clustered under the 'How do I' section of the portal. These services belong to various states/ministries/departments. Since new services are added to the repository of the portal, a need for citizen feedback was felt and, therefore, a rating module was developed in the 'How Do I?' section of the portal. The purpose of introduction of the rating module was to provide a rating to services so that users can assess the services. This would motivate the service provider to improve the service.

The users provide rating to the services based on the two aspects, viz. 'Usefulness' and 'Ease of Access'. Surfers can view the current rating of the service provided by the user. In addition to this, a provision of leaving comments or suggestions by the users has been provided. Visitor's summary and hourly hits by the users have been presented in Tables 6.1 and 6.2, respectively for ready reference.

Universally Accessible
Regardless of the devise being use, technology or ability, NPI is universally accessible to the users. All the initiatives launched are successful. The portal meets priority 1 (level A) of the Web Content Accessibility Guidelines (WCAG) 2.0 placed by the World Wide Web Consortium (W3C).

Table 6.1   Visitor's summary as on 31 March 2017

Visit and Visitors' summary	
Visits	4,371,115
Average per Day	141,003
Average Visit Duration	00:07:35
Median Visit Duration	00:02:02
International Visits	48.64%
Visits of Unknown Origin	0.70%
Visits from Your Country(IN)	50.66%
Visitors	3,697,197
Visitors Who Visited Once	3,414,594
Visitors Who Visited More Than Once	282,603
Average Visits per Visitor	1.18

Source: https://india.gov.in/visitor-summary

Table 6.2   Hourly hits as on 31 March 2017

Hour	Hits
00:00	5,955,712
01:00	3,288,014
02:00	1,984,988
03:00	1,451,618
04:00	1,301,127
05:00	1,593,496
06:00	2,816,841
07:00	5,463,355
08:00	8,507,339
09:00	12,795,250
10:00	19,174,835
11:00	23,977,168
12:00	26,311,450
13:00	23,305,704
14:00	22,029,505
15:00	22,925,496
16:00	22,913,873
17:00	21,670,927
18:00	18,778,846
19:00	17,711,711
20:00	17,203,227
21:00	15,743,976
22:00	13,792,919
23:00	10,177,116

Source: https://india.gov.in/visitor-summary

Customization

Surfers can now choose from a select array of colorful theme options to lend the Home page of National Portal of India a unique, personalized touch. Each theme comes with its own specially designed banner and background data display options.

Nurturing New Initiatives

The National Portal of India has ensured citizen participation in building good governance. The portal-initiated programs such as NGO Partnership System, the Central Information Commission (CIC) Online and Public Opinion on National Testing Scheme (NTS) to solicit public opinions.

Awardees Repository

The 'My India, My Pride' segment of the portal provides search-based access to its Repository of Awardees. This list includes the citizens of India who have received various awards for their devoted service to the nation since the year 1954. NPI has presented the names of receivers of Gallantry Awards and of all brave children who have displayed their courage and have been honored for it.

Guidelines for Indian Government Websites

The National Portal of India was given the responsibility of introducing standards for government's other websites and portals. Central Secretariat Manual of Office Procedures (CSMOP) by the Department of Administrative Reforms and Public Grievances (DARPG) has introduced and published guidelines to be followed by Indian Government Websites (GIGW).

6.3.1.3 Case Summary

The National Portal of India is an online portal that provides a single point access to the different information resources and online services provided by different government sources. The platform has witnessed widespread usage due to its ease of accessibility, easy interactivity and ongoing activities undertaken.

The statistics reveal the following information—NPI had more than 80,000 registered users in 2011 with a growth of 10 percent every month but in recent years, no registration is required for the users to avail services. The portal is gaining heights of popularity and visibility that can be validated from the number of hits it receives in a day. Currently, approximately 7300 forms, 6300 documents, 1900 services, and 1400 schemes are working under NPI. Thus, the portal is expected to witness upward growth in the coming years as well.

6.3.2   Aadhaar Card

6.3.2.1 Overview

Aadhaar is a unique 12-digit number provided to all Indian citizens using biometric inputs. The Aadhaar card number works as identity verification for the citizens and it also helps them avail government services easily. The use of biometric identification process makes the procedure transparent as it reduces the possibility of fake documents. It allows access to government benefits to the areas that do not have banking facilities. For availing banking services like fund transfer, now Aadhaar number has replaced the need of account number, IFSC code. It is more convenient to use Aadhaar number than providing all details like account number, IFSC code and so on as the possession of Aadhaar card establishes a unique identity of the holder.

The Aadhaar Act 2016 established a statutory authority called the Unique Identification Authority of India (UIDAI) meant for Targeted Delivery of Financial and Other Subsidies, Benefits and Services. This comes under the purview of the Ministry of Electronics and Information Technology (MeitY). UIDAI was a part of the then Planning Commission (now NITI Aayog) prior to it being a statutory authority. It was in September 2015 when the government decided to attach the UIDAI to the Department of Electronics and Information Technology (DeitY) of the then Ministry of Communications and Information Technology.

The objective of UIDAI was to allocate Unique Identification Numbers (UID), referred to as "Aadhaar", to the citizens of India. The Unique Identification Numbers has the capability to reduce duplications and eliminate fake identities. The verification can be done in an authenticated, easy and cost-effective manner. The first UID number was allotted to a resident of Nan durbar, Maharashtra, on 29 September 2010. In March 2017, almost 107 crore (more than 1 billion) Aadhaar numbers have been issued by the Authority.

UIDAI is responsible for Aadhaar issuance, authentication and management. UIDAI, under the Aadhaar Act 2016, develops policy, procedure and system for allotting Aadhaar numbers to the citizens. It is also responsible for authentication and security of identity information.

6.3.2.1.1 Vision

The vision of UIDAI is to issue a unique identity to the residents of India to empower them and to provide a digital platform for authentication of the issued anytime, anywhere.

6.3.2.1.2 Mission
The mission of UIDAI includes

(i) Following quality metrics and maintain well-defined turnaround time while issuing Aadhaar numbers to residents.

(ii) Infrastructure-supporting authentication and upgradation of documents need to be developed for the convenience of the residents.

(iii) Team up with other service providers to serve the residents better and equitably.

(iv) A public-private partnership is encouraged for enhancing innovation and to develop Aadhaar-linked applications.

(v) The authority aims to ensure availability and enhancement of the technology infrastructure.

(vi) To ensure that the vision and values of UIDAI are carried forward and build a long-term sustainable organization for that purpose.

(vii) To put forward the work to the best global expertise to invite valuable insights to the UIDAI organization and to enhance the possibility of collaboration.

6.3.2.2 Performance Evaluators

6.3.2.2.1 Number of Aadhaar Cards Issued
The number of enrollments of Aadhaar has doubled since the service has been introduced in 2013. In 2013, the number of enrollments in Aadhaar was 50 crore while in 2016, 100.5 crore have Aadhaar card issued in their name (Figs. 6.2, 6.3 and 6.4).

6.3.2.2.2 Reviewing Aadhaar
The software-based Aadhaar system goes through some stages of typical software development which is called Software Development Life Cycle (SDLC).

Phases of Development
The main idea to study the phases in development is to understand the intermediate deliverables in the process of development of Aadhaar. The inherent part of the process reveals whether the choices and decisions made are in line with the expected outcome or not. If not, a corrective action needs to be taken.

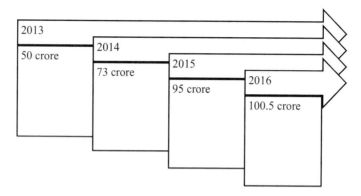

Fig. 6.2 Numbers of enrollments. (Source: Digital India 2016a, b, c)

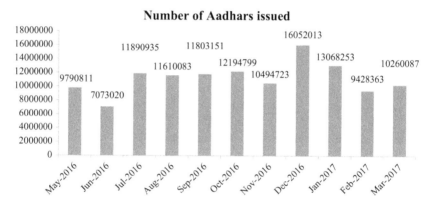

Fig. 6.3 Number of Aadhaars issued in May–December 2016 and January–March 2017 (data as in May–December 2016 and January–March 2017). (Source: https://portal.uidai.gov.in/uidwebportal/dashboard.do?lc=h)

The system planning phase includes project initiation, resource planning, risk analysis and deliverables such as project statement, plan or draft for project execution, and other deliverables. The requirement specifications must have general agreement.

The next phase is system analysis which requires documents to be presented. Basic documentation is required to know the identity of the individual. Requirements documentation are of different types such as textual,

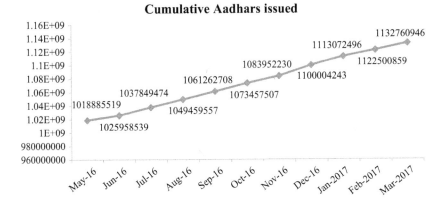

Fig. 6.4 Cumulative Aadhaars issued in May–December 2016 and January–March 2017 (data as in May 2016 and January 2017). (Source: https://portal.uidai.gov.in/uidwebportal/dashboard.do?lc=h)

descriptions, diagrams and prototype displays representing the outcome reports of the developed system. But the requirement documents are negatively known for their inadequacy rather than for their clarity.

Systems design is another phase of development of Aadhaar card. This requires design documents that are the result of the design phase. The outcome of design phase depends on the nature of the project and the methodology followed.

After the system design specification has been made, system implementation is the next step in the process. The system implementation process involves implementation of functioning of complete system. For this, the views of intended beneficiaries of the Aadhaar system, that is, the residents of the country, should understand their requirements. Society at large must accept the government's development agenda which was reflected in the 'Concept Paper on Social Inclusion' issued by the UIDAI in April 2012. The final stage is system support and security that involves operational information system, which includes activities from process optimization to social re-engineering. These new initiatives launched under Aadhaar are results of deliberations and discussions with experts in this domain. Business Process Re-engineering (BPR) was manifested to introduce competition and efficiency into the internal process of Aadhaar. It is an important process in implementation (Rajadhyaksha 2013). The UID project's open architecture has introduced innovative measures. Aadhaar can be used for MGNREGA payments, payment of cooking gas and food grain, cash transfers, and as an identity proof in banks and other transactions. Aadhaar can spur enterprise and consumer applications.

The application of Aadhaar has been beneficial to many industries like banking, insurance, and retail. For example, Aadhaar helps consumer goods companies to supervise the supply chain and enhance integration with retailers in remote areas. Insurance companies are benefitted by Aadhaar application since insurance companies would be needing better information about interested customers. Aadhaar also acts as verification proof in banks, insurance companies and while availing other services. Aadhaar use biometrics, which makes it a reliable service. Biometrics used fingerprints and iris scans to strengthen the project (Punj 2012).

Development of Aadhaar

The new Aadhaar (Targeted Delivery of Financial and other Subsidies, benefits and services) Act, 2016, has been passed to provide subsidies and services to Indian citizens by assigning them Aadhaar numbers.

It is voluntary and not mandatory and can benefit individuals in getting a government service although as per the Supreme Court of India, no person can be denied a service if s/he does not possess Aadhaar card. Aadhaar can be used for Public Distribution System, kerosene and LPG distribution system but for availing these facilities, it is not mandatory to possess an Aadhaar card.

Eligibility

All residents of the country are eligible to acquire an Aadhaar number. The resident of India is a person who has resided in India for at least 182 days in the one year prior the date they apply for the enrollment for Aadhaar.

Information to Be Submitted

An individual, in order to acquire an Aadhaar card, must provide basic information or identity proofs such as his/her biometric (photograph, iris scan, finger print), demographic information (name, date of birth, address) or any other information required by Unique Identification Authority (UID).

Enrollment

For enrollment, an individual will be informed of the way in which the information will be used, the authority or individuals who will receive the information and that they have the right to access their information. After the information provided by a person is verified, Aadhaar number is issued to him/her.

Use of Aadhaar Number

Aadhaar number serves the purpose of an identity proof and is used for receiving a subsidy or a service from the government. In case a person does not have an Aadhaar number, government would require them to obtain it, and temporarily, they will provide him/ her, another means of identification. Aadhaar card is acceptable to all public or private entities as an identity proof for any purpose. However, Aadhaar number cannot be claimed as a proof of citizenship or residence.

Functions and Composition of Authority

An Aadhaar card is issued by the Unique Identification Authority (UID). The UID authority has the following functions.

- To lay down the demographic and biometric information to be collected from the applicants for enrollment,
- To allocate Aadhaar numbers to the applicants,
- To verify Aadhaar numbers, and
- To lay down the usage of Aadhaar numbers for receiving subsidies and services.

The UID authority comprises of a chairperson, two part-time members and a chief executive officer. The chairperson and members of the authority must have an experience of at least ten years in areas like technology, governance, and so on.

Authentication

A facility of authentication through Aadhaar is also provided. An entity or individual can request for authentication of the Aadhaar number of an individual to the Unique Identification Authority (UID). The requesting entity, which can be an agency or a person, is required to obtain the consent of the individual he wants to enquire about, to authenticate information of a person. However, the information provided can only be used for the purposes for which the individual has given consent (Tables 6.3 and 6.4).

6.3.2.3 Case Summary

The Aadhaar program is one among the many success stories of the country and has been observed as an initiative unparalleled in scope anywhere else in the world. Further, the implementation of a national identification program linked with the biometric data is one of its striking features. It has seen a steep rise in the number of registrations for last few years and vast coverage of the population.

Table 6.3   Number of Aadhaar cards issued as on 31 March 2017

Rank	Country/State/Union Territory	Population	AADHAARs issued
	India	1,210,601,445	1,070,261,870
18	Delhi	16,753,235	20,254,032
11	Telangana	35,286,757	37,759,194
15	Punjab	27,704,236	29,327,344
27	Chandigarh	1,054,686	1,110,557
21	Himachal Pradesh	6,856,509	7,206,101
16	Haryana	25,753,081	26,941,701
13	Kerala	33,387,677	34,397,264
26	Pondicherry	1,244,464	1,271,084
10	Andhra Pradesh	49,386,799	50,401,993
32	Andaman and Nicobar Islands	379,944	386,799
17	Chhattisgarh	25,540,196	25,934,547
36	Lakshadweep	64,429	65,300
25	Goa	1,457,723	1,471,613
14	Jharkhand	32,966,238	33,072,792
33	Dadra and Nagar Haveli	342,853	341,829
22	Tripura	3,671,032	3,611,205
2	Maharashtra	112,372,972	109,190,323
30	Sikkim	607,688	590,453
8	Karnataka	61,130,704	57,905,423
5	Madhya Pradesh	72,597,565	68,701,388
19	Uttarakhand	10,116,752	9,453,407
6	Tamil Nadu	72,138,958	65,497,220
9	Gujarat	60,383,628	54,348,784
7	Rajasthan	68,621,012	60,969,794

Source: https://portal.uidai.gov.in/

Table 6.4 Expenditure on Aadhaar cards (2009–2017)

Fiscal year	Expenditure
2009–2010	26.21 Crore (US$3.9 million)
2010–2011	268.41 Crore (US$40 million)
2011–2012	1187.50 Crore (US$180 million)
2012–2013	1338.72 Crore (US$200 million)
2013–2014	1544.44 Crore (US$230 million)
2014–2015	1615.34 Crore (US$240 million)
2015–2016	2039.64 Crore (US $297 million)
2016–2017	8,77,200 Crore (US$140 million)

Source: https://portal.uidai.gov.in/

Moreover, Delhi, Telangana and Haryana have witnessed the maximum number of adoptions. However, the adoption lags in the northastern states. The card is meant to streamline the bureaucratic processes to accelerate the various governance activities. Thus, the ease offered by linking the various services to Aadhaar card has leveraged the convenience levels decreasing the idle time.

6.3.3   Mobile Seva Scheme

6.3.3.1 Overview
The Indian economy has undergone numerous changes post-liberalization. The evident technological paradigm shift and increased internet penetration has resulted in exponential increase in the usage of the different smart devices.

Moreover, the smart phone industry has taken the market by storm. The year 2016 recorded about 684.1 million mobile users, which are estimated to showcase an incline and reach up to 730.7 million users for the year 2017 (Statista Portal 2017).

With the growing concerns for *data transparency* and prevention of *data malware*, a rise in the penetration of different authenticated modes of mobile-equipped services has observed an upsurge. In addition, the Ministry of Electronics and information Technology conceptualized and launched the mobile seva/mobile governance (M-governance), with Centre for Development of Advanced Computing (CDAC) acting as the technical implementing agency (Fig. 6.5).

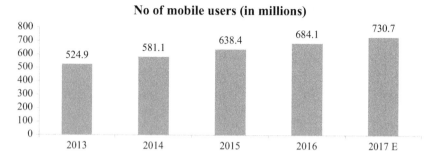

Fig. 6.5 Number of mobile users in India (2013-2017E as on March, 2017). (Source: Statista Portal, https://www.statista.com/statistics/274658/forecast-of-mobile-phone-users-in-india/)

The overall objective of the scheme is to provide an integrated platform to all the Government Departments and Agencies to provide different public services to the citizens and businesses over the mobile devices, using facilities such as SMS (Short Messaging Service), USSD (Unstructured Supplementary Data Service), IVRS, Core Banking Solutions (CBS), and Location-Based Service (LBS).

6.3.3.2 Performance Evaluators

6.3.3.2.1 Mobile Seva: Current Status (Table 6.5)

6.3.3.2.2 Increase in the Number of the Push SMSes

The push messages or the WAP-Push are the messages that comprise of a URL/link in the content to be followed by the user for an action to be initiated or complete a desired task (Table 6.6).

The past performance of this application clearly depicts a rise in the number of Push SMSes for the period 2012–2017. This implies the increase in the number of applications downloaded using the M-App store which points towards the rise in the users for m-governance. Furthermore, the overall increase indicates an optimistic outlook for the usage of the mobile seva scheme.

6.3.3.2.3 Increase in the Number of Pull Messages

Pull messages are the messages which require an initiation to be made from the client's end. The message is initiated by making a request with the help of the desired codes (Table 6.7).

Table 6.5 Current status of mobile seva scheme as on 31 March 2017

Features	Collaborated with	Transactions
Push SMS integration	3609 Departments/Agencies	15,277,640,164
Pull SMS integration	729 Services	17,570,734
IVRS Integration	–	6,073,515
USSD Integration	–	1,386,369
M-app store Status		
Mobile App Development	1025 Live mobile applications (m-Apps)	4,558,009
Number of application downloads	–	3,836,457
Number of Demo applications	–	63

Source: https://mgov.gov.in/index.jsp

Table 6.6 Number of Push SMSes per year and their percentage change as on 1 March 2017

Year	Total Push SMSes per year (in crores)	% change
2012	17,097,598	–
2013	5,534,778,099	32,271.67%
2014	1,677,302,387	(69.70)%
2015	4,878,507,737	190.85%
2016	5,521,235,234	13.17%
2017	2,630,256,216	(52.36)%

Source: https://mgov.gov.in/yearlyGraph.jsp

The scheme recorded a total number of 13,562,552 pull message requests up to March 2017 and offered a total of 695 services to the citizens. Further, the central government recorded the maximum number of message requests of 3,872,192 and the least was accounted by the state of Uttar Pradesh. This implies that the citizens themselves availed the different services available with mobile seva project. In the coming years, the project is expected to showcase a healthy outlook. As the rise in the number of pull message indicates an increase in the number of users which corresponds to a rise in the acceptability towards the application.

6.3.3.2.4 Increase in the Number of Mobile Applications Downloaded via the M-App Store

A centralized platform called *Mobile Service Delivery Gateway* (MSDG) was created to facilitate smooth interaction and transactions across the different integrated state/central government departments or agencies (Table 6.8).

Table 6.7   Total number of requests and services made for the pull messages

S No.	Service type (Central/State/UT)	Total no. of request	Total no. of services
1	Central	3,872,192	110
2	PSU (Central and State)	10,504	7
3	Andaman and Nicobar	17,035	16
4	Andhra Pradesh	467,684	24
5	Arunachal Pradesh	113	3
6	Bihar	18,419	16
7	Chandigarh	6734	10
8	Chhattisgarh	1322	3
9	Dadra and Nagar Haveli	593	75
10	Daman and Diu	404	67
11	Delhi	3911	2
12	Goa	208,565	5
13	Gujarat	29,296	12
14	Haryana	19,419	15
15	Himachal Pradesh	171,798	22
16	Jharkhand	3,268,090	14
17	Karnataka	73,809	6
18	Kerala	483,764	9
19	Madhya Pradesh	76,150	17
20	Maharashtra	1,448,378	46
21	Manipur	2562	14
22	Meghalaya	188	2
23	Mizoram	54,694	17
24	Nagaland	4503	26
25	Odisha	6932	15
26	Punjab	1,126,361	11
27	Rajasthan	592,396	23
28	Sikkim	1473	48
29	Tamil Nadu	261	3
30	Telangana	695	2
31	Tripura	2286	3
32	Uttar Pradesh	134	2
33	Uttaranchal	247	2
34	West Bengal	900,461	48
35	Invalid Requests	676,274	0
	Total request:	13,562,552	695

In addition, there are about 1087 applications available with the m-app store and a total of 3,836,430 downloads have been made so far. The various domains, this scheme encompasses are AADHAAR, Agriculture, Electoral, Health, Indian Post, judiciary, Language, M-learning, Transport and many others. Further, it comprises of the various utility applications as well which includes My MTNL Delhi, HP Gas, M-Gov Appstore, SCM AP.

Table 6.8 State-wise download of applications available in the m-app store as on 1 March 2017

S. No.	Service type (Central/State/UT)	Total no. of applications	Total no. of downloads
1	Nagaland	1	794
2	Mizoram	2	38
3	Tamil Nadu	2	699
4	Uttar Pradesh	3	28,807
5	Delhi	4	1101
6	Arunachal Pradesh	4	1327
7	Karnataka	4	4021
8	Chandigarh	4	4552
9	Chhattisgarh	4	6825
10	Meghalaya	6	763
11	Haryana	7	3199
12	Gujarat	10	1129
13	Jharkhand	10	27,901
14	Kerala	11	917
15	Goa	11	3464
16	Manipur	11	5751
17	Odisha	12	451
18	Bihar	12	8037
19	Madhya Pradesh	13	12,952
20	Punjab	20	18,629
21	Andhra Pradesh	26	15,577
22	West Bengal	27	11,968
23	Rajasthan	30	24,621
24	Others	31	20,580
25	Himachal Pradesh	35	12,585
26	Sikkim	61	30,418
27	Maharashtra	68	26,125
28	General Application	126	27,136
29	Central Government Service	509	315,915
30	Demo Application	63	19,600

Source: www.mgov.gov.in

Table 6.9 represents the number of mobile applications available with the M-App store for the different states and the number of downloads done. Thereby, Mizoram recorded the least number of downloads of 38 and Central Government accounted for the maximum downloads of 315,915. In addition, the demo application witnessed a download of 19,600 times, which expresses the likeability towards the initiative (Figs. 6.6 and 6.7).

Table 6.9   Government initiatives to enhance E-Participation

Beti Bachao Beti Padao	Chemicals and petrochemicals	Creative corner
Department of Administrative Reforms and Public Grievances	Digital India	Energy Conservation
Girl Child Education	Healthy India	Indian Railways
Manual Scavenging Free India	Ministry of Defense	MSME
New Education Policy	Open Forum	Sakriya Panchayat
Sporty India	Watershed Management	Caring for the Specially abled
Clean Ganga	Dadra Nagar Haveli UT	Department of Industrial Policy and Promotion (DIPP)
Disaster Resilient India	Expenditure Management Commission	Government Advertisements
Incredible India	Job Creation	Ministry of Agriculture and Farmer Welfare
Ministry of Housing and Urban Poverty Alleviation	Ministry of Petroleum and Natural Gas	Ministry of Steel
NITI Aayog	Rural Development	Skill Development
Swachh Bharat	Youth for Nation Building	Chandigarh UT
Consumer Protection and Internal Trade	Daman and Diu UT	Department of Revenue
Department of Telecom	Economic Affairs	Food Security
Green India	India Textiles	Mann ki Baat
Ministry of coal	Ministry of culture	Ministry of Human Resource Development (MHRD)
Ministry of Power	MyGov Move-Volunteer	NRI's for India's growth
Saansad Adarsh Gram Yojana	Smart Cities	Tribal Development

Source: MyGov (2016a, b)

6.3.3.2.5   Awards and Recognitions

2013

The scheme coveted the titles of m-Billionth South Asia award for the year 2013 wherein the award recognizes the initiatives made in the domain of technology which support higher levels of innovation.

2014

For the year 2014, the scheme was entitled with the "United Nations Public Service Award" for India's mobile governance initiative, under the category of "Promoting Whole of Government Approach in the Information Age" (mgov.in 2017).

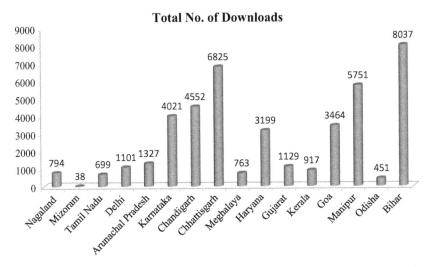

Fig. 6.6 State-wise download of the applications available in the m-appstore (in crores) as on 1 March 2017. (Source: mgov.in)

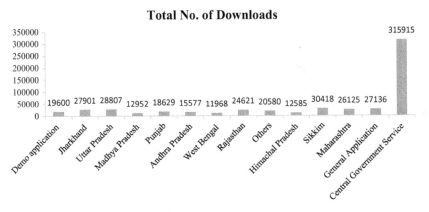

Fig. 6.7 State-wise download of the applications available in the M-Appstore (in crores) as on 1 March 2017. (Source: www.mgov.gov.in)

6.3.3.3 Case Summary

Mobile seva is one such initiative in the domain of technology which aims at increasing the transparency levels of the different transactions being carried out among the citizens and different departments and agencies integrated with m-governance.

This scheme is one among the numerous E-Governance initiatives which emphasizes on the accountability and fairness parameters for conducting the different actions linked with the different government departments. Moreover, the scheme has exhibited a positive growth since its date of inception and is estimated to yield a positive response in the coming years as well.

6.3.4   MyGov.in

6.3.4.1   Overview

Mygov.in is a platform wherein the citizens have the liberty to actively participate in the nation's governance and development. Further, the citizens have the right to discuss and contribute to the plans of government with the help of this platform.

Moreover, MyGov is one such initiative undertaken by the government to augment the E-Participation of the citizens (Table 6.10).

MyGov.in aims at bringing the government closer to the common man by encouraging the public to use online platform for sharing their ideas and opinions about government initiatives. This platform has been used to encourage participation of general public and experts to exchange ideas and views with the government to facilitate social and economic transformation of India. Since its inception, the platform of MyGov has succeeded to keep the citizens involved in decision making regarding governance initiatives such as Girl Child Education, Skill Development, Rural Development, Clean Ganga, Green India and so on. This platform has made all possible efforts to bridge the gap between the citizens and the government, which was traditionally very large.

Apart from these initiatives, the government also enhanced the contribution of citizens by encouraging them to participate in decision-making and soliciting suggestions regarding logos, taglines, slogans, visual designs,

Table 6.10   Status for MyGov.in statistics as in March 2017

Registered number of Users	4.17 Crore
Number of tasks performed	612
Number of submissions	18,699,000
Number of discussions made	714
Number of comments delivered	368,727,000

Source: https://www.mygov.in/

and so on. To enhance the participation of citizens, people are invited to design logos, themes, contests, posters, competition to make their city smart. Suggestions are invited from citizens through polls and surveys about various schemes and issues. Smart city proposals, draft for procurement and conservation, mega sports events, elections and proposals on schemes, regulations and processes are discussed with the public and their opinions are solicited. The blogs under MyGov talk about the successful events, the contribution of agencies like Council of Scientific and Industrial Research (CSIR) is appreciated (MyGov.in 2017).

The final section involves talks between the government and the public regarding the propositions. Various seminars, panel discussions, and radio programs are organized to know the feedback of general public.

6.3.4.2 Performance Evaluators

6.3.4.2.1 User Participation

The very purpose of MyGov is to bring the government closer to the common man by encouraging the public to use online platform for sharing their ideas and opinions about government initiatives. This platform is used to encourage participation of general public and experts to exchange ideas and views with the government to facilitate social and economic transformation of India. Since its inception, the platform of MyGov has succeeded to keep the citizens involved in decision making regarding governance initiatives such as Girl Child Education, Skill Development, Rural Development, Clean Ganga, Green India, and so on. This platform has made all possible efforts to bridge the gap between the citizens and the government, which was traditionally very large. Today, MyGov has more than 1.78 million users who share their thoughts through discussions and also contribute towards assigned tasks. Also, this platform receives more than 10,000 posts per week in the form of suggestions; these suggestions are then delivered to the concerned departments who have the capability to transform them into actionable agenda (MyGov 2016a, b).

6.3.4.2.2 Decision Making

MyGov.in is a platform for the citizens to contribute to the decision-making process in governance of the country. The citizens are made a part of the policy formulation and their suggestions are also taken into consideration while making important governance decisions. Since this platform is quite important in highlighting the role of citizens and recognizing their contribution in governance to transform India, the platform undergoes upgradations in order to improve user experience. The major features of MyGov are discussions, talks, polls, tasks, surveys and blogs on public policy and governance issues (MyGov 2016a, b).

6.3.4.2.3 Digital Take Up

The platform has experienced impressive numbers and has a long way to go with this progress. As Honorable Prime Minister of India, Shri Narendra Modi, quoted that MyGov is a means to initiate a mass movement towards self-governance or 'Surajya'. MyGov takes the digital way up to collect citizens' inputs who are the ultimate beneficiaries of the progress made and utilize them to implement policy change. MyGov is now progressing under the Digital India campaign and utilizing the technology to further enhance the public service delivery and citizen engagement (Alawadhi 2016).

6.3.4.2.4 Improvement Initiatives

Promoting the belief that the government is of the people, by the people and for the people, the government of India has provided a platform for the citizens to share their ideas and suggestions for the development of the nation. MyGov is based on the principle of 'Discuss, Do and Disseminate' providing opportunity to the citizens to share their views and discuss their ideas with domain experts on national issues. Polls and tasks are other facilities provided to facilitate nation building.

Blogs

Blogs are written by municipal commissioners of cities thanking citizens for their participation and recommendations for improvement.

Contests

MyGov contests are meant to recognize Citizen Brand ambassadors of various cities. These ambassadors carry the baggage of promoting aspects of citizenship and come as a leader to facilitate negotiation of implementation issues with residents.

Closed Group

Closed groups have been formed for discussions of ideas and procedure for implementation of programs. These groups involve municipal commissioners of winning cities, experts and Ministry of Urban Development (MoUD) officials.

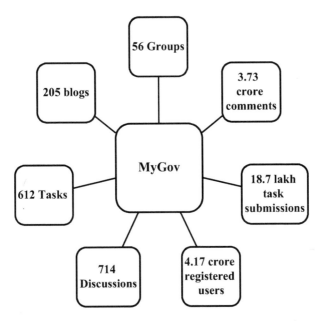

Fig. 6.8 MyGov statistics (data as on 28 April 2017). (Source: https://www.mygov.in/)

Talk
This involves talks by commissioners on a quarterly basis. The aim of these talks is to update citizens on progress taking place and soliciting questions from them to enhance transparency in the process (Figs. 6.8).

MyGov Volunteer App
MyGov Volunteer app can be used by the citizens to engage themselves with the various volunteer activities (Figs. 6.9).

6.3.4.2.5 Citizens' Reviews and Suggestions

Ashish and Gaurav, Residents of Delhi Share Their Opinions for the Platform

> *MyGov app in my opinion best app by Government. The app simply provided a way to participate in nation building progress and a platform where I can directly share my view and ideas with the Government. Proud to be a part of this app. (Digital India 2016a, b, c)*

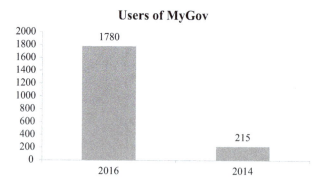

Fig. 6.9 Registered users of MyGov (data as in November 2016). (Source: Digital India 2016a, b, c)

> *Our government must implement TRUE net neutrality, not a PSEUDO net neutrality as defined by telecom companies. All apps/websites should be treated equally and no preference should be given to any app/website. No privilege should be given to a website/app that pays TRAI to have users browse its content for free. Citizens of India have already struggled a lot in the earlier License Raj. The government should not go back to the License Raj, by avoiding licensing for services like Whatsapp, Skype. (Digital India 2016a, b, c)*

6.3.4.3 Case Summary

Just after 45 days of its launch, MyGov.in was able to attract participation of 215,000 citizens who were enrolled to MyGov. More than 28,000 users contributed ideas and plans about various issues just within 45 days of launch of MyGov. The Prime Minister adopted ideas submitted by general public to MyGov to develop his online Independence Day message.

The platform also has attracted many users who were not involved in other governmental social media accounts such as Facebook and Twitter. More than 23,000 entries for 7 different government Ministries were received within 50 days of inception. These were in addition to the Prime Minister's Office and these entries came alone through the Creative Corner section (Alawadhi 2016).

6.3.5   E-Transaction Aggregation and Analysis Layer (E-Taal)

6.3.5.1 Overview

Since lately, the Indian economy has undergone a technological paradigm shift. With the change in the standard of living and rising government initiatives such as Digital India, Bharat Net, E-Kranti, Smart City initiative, an evident penetration of the different smart devices has been observed. Furthermore, the recent years have witnessed an increased usage of the internet-enabled devices to facilitate an ease of performing transactions on different platforms. In addition, the framework of E-Governance has become more robust with the help of the emerging concepts and schemes revolving around the parameters of fairness, accountability and data transparency.

Thus, in support of the E-Governance initiative, a web-based application called E-Taal was developed and launched by the National Informatics Centre (NIC). Further, E-Taal is a mechanism that allows the consolidation, analysis of the delivery of services available at the different E-Governance projects. The application allows analyzing the impact of different electronic governance programs at both the national and state levels. In addition, E-Taal is an electronic transaction service that uses the different Information and Communication Technology tools and platforms to provide quick accessibility and reduce the query response time.

The project is based on the principle of *"You can MANAGE effectively, what you can MEASURE"* (Table 6.11).

6.3.5.2 Key Performance Indicators

6.3.5.2.1 Number of Transactions Performed Using E-Taal

E-Taal gives a graphical representation of the different transaction counts done by different E-Governance projects. Further, E-Taal provides a transaction count of the various E-Services made available to its citizens across the nation. The scheme caters about 3267 E-Services which include data from *21 Central Ministries, 36 states* and *Union Territories* (Table 6.12).

This implies that the scheme has observed a nationwide acceptability since its inception in January 2013. Moreover, the period 2013–2017 has witnessed an increase in the number of transactions. Therefore, the project is estimated to showcase an increase in the number of transactions in the coming years too.

Table 6.11 E-Taal: An overview (data as accessed on 1 April 2017)

Number of E-Services integrated	3267
National E-Transaction count (1 January 2017)	1,668,066,809
National E-Transaction count (1 April 2017)	920,517,341

Source: http://etaal.gov.in/etaal/auth/login.aspx

Table 6.12 Total number of transactions done using E-Taal (2013–2017) (data as accessed on 18 February 2017)

Year	Number of E-Transaction	Number of average transactions
2013	2,417,658,057	6,468,644
2014	3,576,949,576	9,642,707
2015	7,607,536,283	20,680,918
2016	10,887,563,104	29,585,299
2017 (*till February*)	1,668,068,899	34,728,415

Source: http://etaal.gov.in/etaal/auth/login.aspx

Table 6.13 E-Taal Performance up to February 2017

Top 5 States of February 2017	Gujarat, Andhra Pradesh, Telangana, Tamil Nadu, Madhya Pradesh
Top 2 Union Territories of February 2017	Delhi, Chandigarh
Top 5 Central Projects of February 2017	Unique Identification Authority of India (UIDAI) Services, Agriculture, Public Financial Management System, Railway Reservations, Judiciary
Top 5 State MMPs of February 2017	Land Records, PDS, Commercial Taxes, E-District, Treasuries
Top 5 Central MMPs of February 2017	Agriculture, E-Courts, Passport, Income Tax (IT), CSC
Top 5 Standard Services of February 2017	Agriculture and Allied, Social Welfare and Pension, Rural Development, Land Revenue, Public Distribution System

Sources: http://etaal.gov.in/etaal/auth/login.aspx and etaal.gov.in

6.3.5.2.2 Integration with the Various States, Mission Mode Projects (MMPs) and Standard Services

Currently, the scheme covers about 19 MMPs namely *Income Tax (IT), MCA 21, Passport, Immigration, Visa and Foreigners Registration and Tracking (IVFRT), Pension, Agriculture, Commercial Taxes, E-District, Employment Exchange, Land records, Municipalities, Gram Panchayats,* which have been integrated with 'E-Taal' (Tables 6.13 and 6.14).

It can be clearly seen that several union ministries have integrated their transactions with E-Taal. This directly correlates the data transparency with the ease of accessing the data. In addition, Ministry of Electronics and Information Technology recorded the highest number of E-Transactions followed by Ministry of Agriculture and Farmers Welfare and Ministry of Railways.

Table 6.14 Union Ministries and the number of transactions performed (from January 2017 to February 2017)

Union Ministries	No. of E-Transactions	No. of E-Services
Ministry of Agriculture and Farmers Welfare	207,977,812	3
Ministry of Commerce and Industry	23,946	1
Ministry of Communications	5,883,944	1
Ministry of Consumer Affairs, Food and Public Distribution	738,198	7
Ministry of Corporate Affairs	471,965	45
Ministry of Electronics and Information Technology	1,052,357,904	100
Ministry of External Affairs	15,289,117	75
Ministry of Finance	57,347,239	14
Ministry of Health and Family Welfare	3,874,036	9
Ministry of Home Affairs	15,425,424	4
Ministry of Labour and Employment	6811	1
Ministry of Law and Justice	39,394,174	7
Ministry of Panchayati Raj	496,798	38
Ministry of Personnel, Public Grievances and Pensions	3,895,121	13
Ministry of Railways	65,172,824	4
Ministry of Road Transport and Highways	267,787	13
Ministry of Rural Development	3,299,231	2
Ministry of Urban Development	230,550	1
Total E-Transactions	**1,472,152,881**	

Source: etaal.gov.in

6.3.5.2.3 Center-wise Transactions

The center-wise transactions represent that the maximum number of E-Transactions was carried out by Unique Identification Authority of India (UIDAI) Services, followed by Kissan SMS portal and Railway Reservations (Table 6.15).

6.3.5.2.4 State-wise Transactions

The top five states which have integrated their transactions with E-Taal are Gujarat, Andhra Pradesh, Telangana, Tamil Nadu, and Madhya Pradesh. Andhra Pradesh has accounted for about 59,112,401 transactions followed by Gujarat which registered about 56,591,470 transactions (Table 6.16).

Table 6.15 The Central-wise report (January 2017 to April 2017)

SL	Project	No. of services	No. of transactions	E-Transaction per services	E-Transaction % share
1	AGMARKNET	1	120,240	120,240	0.01
2	Central Public Works Department (CPWD)	1	240,189	240,189	0.01
3	Centralized Visa Issuance System	29	420,398	14,496.48	0.02
4	Centralized Public Grievance Redress And Monitoring System (CPGRAMS)	1	1,532,732	1,532,732	0.09
5	CONFONET	4	730,888	182,722	0.04
6	DAVP-Advertisement	2	53,210	26,605	0
7	DAVP-Empanelment	3	593	197.67	0
8	Dial.gov	1	1062	1062	0
9	ECI—National Electoral Services	4	15,823,604	3,955,901	0.88
10	E-Courts	7	43,415,118	6,202,159.71	2.42
11	Employment Exchange	1	6950	6950	0
12	eSangam	1	1,814,643	1,814,643	0.1
13	Farmers Web Registration	1	137,599	137,599	0.01
14	FORM C Services	1	2,099,215	2,099,215	0.12
15	FORM S Services	1	3432	3432	0
16	Identity Card Management System	1	1196	1196	0
17	Interactive Information Dissemination System (IIDS)	2	1,615,886	807,943	0.09
18	Jeevan Pramaan: Life Certificate for Pensioners	1	1,276,820	1,276,820	0.07
19	Kisaan SMS Portal	1	209,920,176	209,920,176	11.68
20	MCA21	45	508,112	11,291.38	0.03
21	NCCPR of TRAI	1	6,311,501	6,311,501	0.35
22	Nikshay	3	252,572	84,190.67	0.01
23	No Obligation to Return India	1	162	162	0
24	Online FRRO and FRO Services	20	72,935	3646.75	0
25	Overseas Citizenship of India (OCI)	6	802,680	133,780	0.04

(*continued*)

Table 6.15 (continued)

SL	Project	No. of services	No. of transactions	E-Transaction per services	E-Transaction % share
26	Passport	17	13,202,108	776,594.59	0.73
27	Pensioners Portal	2	10,194	5097	0
28	Public Financial Management System	7	49,923,294	7,131,899.14	2.78
29	Railway Reservations	4	68,671,015	17,167,753.75	3.82
30	Right To Information	4	6227	1556.75	0
31	TAX INFORMATION NETWORK	2	7,213,449	3,606,724.5	0.4
32	Unique Identification Authority of India (UIDAI) Services	2	1,369,879,991	684,939,995.5	76.22
33	UPSC	5	1,314,030	262,806	0.07
	Total	182	1,797,382,221	948,781,278	

Source: http://etaal.gov.in/etaal/PopReportCustom.aspx

6.3.5.2.5 Awards and Recognitions (in Chronological Order)

2016

In the Digital India Knowledge Exchange Summit, held in Kerala, E-Taal was awarded as the *Best Digital India Initiative.*

2014

The E-Taal project was adjudged as one of the best projects in *E-INDIA* awards under the G2C (Government to Citizens) awards category.

2013

At the Skoch Digital Inclusion Awards, NIC coveted the esteemed Platinum Award and Order of Merit.

6.3.5.3 Case Summary

E-Taal allows convenient and timely visualization of the data for the numerous E-Governance and Mission Mode Projects. Moreover, the characteristic of E-Taal that helps to identify the number of end-to-end electronic transactions was declared as the best indicator for measuring real-time performance of E-Governance services in terms of service delivery to citizens. The numbers and statistics on E-Taal are growing with each passing year and a number of services/departments are getting registered on the platform.

Table 6.16   State-wise report (January 2017 to February 2017)

SL	State	No. of services	No. of transactions	E-Transaction per services	E-Transaction % share
1	Andaman and Nicobar	9	55,946	6216.22	0
2	Andhra Pradesh	169	71,628,767	423,838.86	3.02
3	Arunachal Pradesh	15	111,487	7432.47	0
4	Assam	43	3,910,351	90,938.4	0.16
5	Bihar	17	8,521,196	501,246.82	0.36
6	Chandigarh	49	803,686	16,401.76	0.03
7	Chhattisgarh	47	21,677,562	461,224.72	0.91
8	Dadra and Nagar Haveli	27	101,806	3770.59	0
9	Daman and Diu	26	114,376	4399.08	0
10	Delhi	50	13,301,323	266,026.46	0.56
11	Goa	30	657,064	21,902.13	0.03
12	Gujarat	130	66,638,607	512,604.67	2.81
13	Haryana	57	15,355,005	269,386.05	0.65
14	Himachal Pradesh	49	6,276,803	128,098.02	0.26
15	Jammu and Kashmir	22	1,615,431	73,428.68	0.07
16	Jharkhand	34	3,179,487	93,514.32	0.13
17	Karnataka	52	15,832,128	304,464	0.67
18	Kerala	140	36,481,311	260,580.79	1.54
19	Lakshadweep	9	130,165	14,462.78	0.01
20	Madhya Pradesh	120	32,196,193	268,301.61	1.36
21	Maharashtra	45	20,791,490	462,033.11	0.88
22	Manipur	23	489,168	21,268.17	0.02
23	Meghalaya	30	1,189,293	39,643.1	0.05
24	Mizoram	31	273,447	8820.87	0.01
25	Nagaland	10	98,789	9878.9	0
26	Odisha	49	8,105,287	165,414.02	0.34
27	Puducherry	35	646,919	18,483.4	0.03
28	Punjab	92	8,983,023	97,641.55	0.38
29	Rajasthan	150	42,751,708	285,011.39	1.8
30	Sikkim	8	115,845	14,480.63	0
31	Tamil Nadu	59	52,260,686	885,774.34	2.2
32	Telangana	124	52,575,037	423,992.23	2.22
33	Tripura	38	594,378	15,641.53	0.03
34	Uttar Pradesh	53	48,172,676	908,918.42	2.03
35	Uttarakhand	34	2,523,808	74,229.65	0.11
36	West Bengal	92	36,847,616	400,517.57	1.55
	Total	1968	575,007,864	7,559,987.31	

Source: http://etaal.gov.in/etaal/PopReportCustom.aspx

Therefore, E-Taal is expected to showcase a healthy growth in the future due to the adaption of the different services in various states and department using upcoming technologies for the need to promote E-Governance. E-Taal will help to develop a robust technological framework for various E-Governance services in the country.

6.3.6   Mann Ki Baat

6.3.6.1 Overview
Mann Ki Baat is a broadcast show headed by Honorable Prime Minister of India, Shri Narendra Modi. Through this radio program, the prime minister addresses the Indian citizens using various platforms like radio, DD national and DD news (*The Hindu* 2014). The prime minister has addressed the audience in 24 Mann ki Baat programs where more than 61,000 ideas have been presented by him and 1.43 lakh audio have been received. A few selected calls become a part of the broadcast each month.

The feeds of 20-minute-long episodes are communicated by Doordarshan's Direct to Home (DTH) service to television and radio channels. The President of the United States participated in an episode of 'Mann Ki Baat' that was aired on 27 January 2015.

The prime minister reaches the general masses of India through All India Radio since 3 October 2014. Radio is chosen as a medium because television or internet is not available in the rural and remote areas of India. Radio has a wider reach than any other medium. Ninety percent of the population can be reached through radio. Apart from All India Radio, private FM radio stations are permitted to broadcast the show. The schemes, ideas and initiatives launched by the government such as Swachh Bharat Abhiyaan, Mars Mission, One Rank One Pension are promoted through the radio show. The prime minister also uses this platform for spreading awareness on various national issues like terrorism, drug abuse, concerns to farmers and so on. 'Mann ki Baat' is a large revenue-generating program for All India Radio since it is attracting a large number of sponsors and advertisers (MyGov 2016a, b).

The purpose of the radio show is to develop a regular connection with the common man, to interact with him about the initiatives and to invite feedback from the people in nation building. This radio program allows the citizens to give advice on areas and topics that the government should focus.

6.3.6.2 *Performance Evaluators*

6.3.6.2.1   Delivery of Successful Episodes on AIR

Mann Ki Baat is a monthly program aired on AIR wherein PM Narendra Modi speaks his heart out. The first addressal made by him was on 3 October 2014 which focused on the successful Mars Mission and also discussed about the Swachh Bharat Mission launched on 2 October 2014 (Table 6.17).

Table 6.17   Topics of Mann ki Baat Episodes

Dates of episodes	Topic
3-Oct-14	Swachh Bharat Abhiyaan, Mars Mission
2-Nov-14	Scholarship for disabled children
14-Dec-14	Drug Abuse
27-Jan-15	Question answer session with President Obama
22-Feb-15	Exam Stress
22-Mar-15	Issues of concern to farmers
26-Apr-15	One rank one pension
31-May-15	Heat wave, Kissan TV channel
28-Jun-15	Selfie with daughter
26-Jul-15	Road safety, social issues
30-Aug-15	Sunday on cycle, Dengue treatment, Infant mortality, Land Acquisition bill, Zero account balance
20-Sep-15	Initiative of Incredible India
25-Oct-15	PM announced no interview in non-gazette government jobs for Group B, C and D posts and gold monetization.
27-Dec-15	Technology and innovation
31-Jan-16	Beti Bachao Beti Padhao, Pradhan Mantri Fasal Beema Yojana
28-Feb-16	Exam stress
27-Mar-16	PM motivates the youth to become ambassador of 2017 FIFA Under-17 World Cup
24-Apr-16	Conservation of water
22-May-16	Environment, Jan Dhan Yojana
26-Jun-16	Importance of Yoga
31-Jul-16	Sports—Tribute to Dhyanchand. Rio Olympics—praised Sindhu, Deepa and encouragement to women in sports
28-Aug-16	PM talks about innocents being used for spreading violence. The PM also talks about his plans for Tokyo 2020 Olympics, border issues, religion and educational reforms.
25-Sep-16	URI Attack—tribute to 18 martyred soldiers. PM announced to take revenge from the culprits

(continued)

Table 6.17   (continued)

Dates of episodes	Topic
30-Oct-16	'Sab Ka Saath, Sab Ka Vikas', PM wished the audience for Diwali and Guru Nanak Dev's birthday
27-Nov-2016	India's commitment towards demonetization
25-Dec-2016	The prime minister invites you to share your ideas on topics he should address
26-Feb-2017	Prime Minister appreciated ISRO's recent achievements

Source: PMINDIA (2016)

Till March 2017, 29 successful episodes have been delivered. The different editions delivered so far have brought to notice myriad of concerns. The prime minister has lauded the Indian Space Research Organization for scripting history to become the first ever nation to launch successfully 104 satellites into space at one go. Further, he has also appreciated the role of the farmers for record production of food grains that year. It was observed that more than 2700 lakh tonnes of food grains were produced in the country which accounted for a comparatively 8 percent more of yield than the year 2016 (*Times of India* 2017).

6.3.6.3   Case Summary

The program has received huge appreciation by a large number of audiences, particularly among the urban masses residing in the metropolitan cities across the nation. Recently, a survey conducted revealed that about 66.7 percent of the total population had tuned into the prime minister's addressal and found it useful. The usual ad slots on AIR sold for are of the price range 500 (US$7.40) to 1500 (US$22) per 10 seconds, but a 10-second ad slot for Mann Ki Baat cost 2 lakh (US$3000) to the advertisers (*Times of India* 2016).

Till date numerous episodes have been aired on the All India Radio and televisions. It has been made available in even different regional languages and have formed the major source of revenue for AIR. The views expressed by Mr. Narendra Modi focus on a range of aspects such as Swachh Bharat mission, Mars mission, reaching the remote areas, differently abled children (divyang), teachings of Mahatma Gandhi, Swami Vivekananda and a lot more. Further, he even encourages the audiences to buy at least one Khadi garment to help those who are entirely dependent on it. Prime Minister Modi also urges the listeners to share their opinions and suggestions via *www.mygov*.in to enhance citizen participation.

In his words, *"if everyone takes a step forward then India will take 125 Crore steps"*. Thus, his idea of growing with his people and inculcate a sense of belonging amongst all would lead to the creation of a more powerful nation. The program "Mann ki Baat" has certainly gained popularity and is slated to be the most iconic citizen engagement activity by Government of India. Lots of people look forward to it and try to contribute through various contests and activities associated with it.

6.3.7   Jeevan Pramaan

6.3.7.1   Overview

Earlier, it was mandatory for the pensioner to be physically present in front of the Pension Disbursing Agency or submit their life certificates to the Central Pension Accounting Office (CPAO) to procure their pensions. However, the scenarios have changed now. It is no more compulsory for the pensioner to be physically present at the time of disbursements.

The Digital Life certification for pensioners also known as Jeevan Pramaan emphasizes on decreasing the unease associated with the cumbersomeness of pension transfer. Jeevan Pramaan scheme is nothing but an Aadhaar platform which utilizes the biometric authentication of the pensioner. A Digital Life Certificate is usually generated upon the successful completion of the authentication process and gets stored in the Life Certificate Repository. The scheme no more requires an individual to go from pillar to post to get the work done; instead, it has reduced the manual intervention to a great extent.

In addition, the certificates can be retrieved easily via electronic means. The pensioner can easily locate a Jeevan Pramaan Center which is being operated by banks, government offices, Common Service Centres or even via the client application on any PC/mobile/Tablet. One of the most crucial perquisites for a pensioner is to provide a life certificate to the pension disbursing agencies so that their pension is credited into their accounts. At times the need to be present in front of the agencies becomes an obstruction in the hassle-free transfer of the amount to the pensioners. It also augments the amount of inconvenience level, particularly for the aged people, who are not always in a position to present themselves before the authorities. Thus, issuing a life certificate will ensure that the pension is being credited to the account every year.

Furthermore, India has about one crore pensioner families, where pension forms a vital source of income and sustenance. It has been found that there exist about 50 lakh pensioners of the Central government and an almost similar number of the various states and UTs. Over 25 lakh retired army personnel rely on the amount received from pension.

6.3.7.2 Performance Evaluators

6.3.7.2.1 Number of Registered Users

The scheme has observed an increase in the number of users since its inception in 2014. It has recorded about 65.83 lakh pensioners spanning all the states across the nation. Keeping in mind the huge population of the country and about a crore of them depends upon their pensions, the scheme still has a long way to go and reach every individual (jeevanpramaan.gov.in 2017).

However, the scheme primarily focuses on digitizing the whole process of obtaining the life certificates which is likely to grow in the coming years owing to the evident technological advancements.

6.3.7.2.2 Reach of Jeevan Pramaan Centers

The scheme can be accessed with the help of different smart devices and an application for this scheme is also available which can be easily downloaded. In addition, the scheme also offers its users the option to locate a center with respect to their current location.

The center can be located with the help of the user's location or pin code. For the location type, there are four possible options such as Citizen Service Center, National Institute of Electronics and Information Center, Government Offices and J&K banks. In addition, a list of the nearby centers can also be retrieved by sending in an SMS in the format JPL <Pin code> at 7738299899.

The reachability of this application is another parameter which clearly defines the success of this platform.

6.3.7.3 Case Summary

Pension forms a crucial part of survival for the retired citizens. With an aim to benefit over crores of pensioners in the country, the scheme works on an Aadhaar-enabled mechanism to facilitate ease to the common man. The numbers are rising for the usage and many citizens are getting benefitted through this scheme.

Further, it will help to eliminate the undue hardships for the pensioners involved in extracting the life certificates. Thus, the scheme is estimated to exhibit an optimistic growth in the coming years too, owing to the innumerable benefits it has to offer to the retired senior strata of the society.

6.3.8   PayGov

6.3.8.1 Overview

India is a developing economy and has witnessed numerous changes post the *Liberalization, Privatization and Globalization* process initiated in 1990. Due to the evident technological paradigm shift coupled with the government initiatives, a widespread inclination towards the different electronic modes is escalating. Moreover, a rise in the disposable income and growth in the number of paperless transactions is another factor in its support. In addition, the E-Commerce industry is taking the market by storm, which is estimated to grow manifolds in the coming years as well. Thus, arises the need to carry out hassle-free and quick online transactions.

The *National Securities Depository Limited* (NSDL) in collaboration with the *Department of Electronics and Information Technology* (DeitY) and the *Government of India* has led to the formation of a centralized platform to perform the collection of the online payments from the citizens for government services.

PayGov focuses on:

1. Acting as an auxiliary for the E-Governance initiative;
2. Online tender procurement;
3. Conducting conferences;
4. Admissions to universities, medical institutes and
5. Collection of relief fund donation

6.3.8.1.1 Objectives (Fig. 6.10)

6.3.8.1.2 Features

PayGov is a centralized *online payment gateway solution*. The platform can be integrated with the online service portal of the department willing to use this architecture. Furthermore, PayGov comprises of certain predefined set of standards to be followed, coupled with the procedures and a set of protocol to facilitate ease of communication upon its deployment.

To provide *One Platform for Payments* and *Settlement* solution

This facilitates the ease of carrying out the desired payments via citizens debiting the amount from their banks, etc. and settlement of monies with all the settled tax including the Basic Statistical Returns along with Corporate identity Number of the agency. Furthermore, in the case of a treasury payment, it is usually settled in T+1 days.

Generation of MIS & reports:

Management Information System reports that allow in keeping a track of the day-to-day business activities are generate in accordance with the requirement and the format pre-defined by the Government portal.

Formulation of a unified process coupled with the treatment of broken transactions

In accordance with the policies the Government department portals can either auto settle or auto cancel as per the advice of the Government department with the help of an automated process.

Fig. 6.10 Objectives of PayGov. (Source: http://paygovindia.gov.in/)

In addition, PayGov is compatible with *National E-Governance Services Delivery Gateway (NGSDG)* and *State E-Governance Services Delivery Gateway (SGSDG)*. Moreover, the integration involves carrying out certain paper formalities for successful integration.

PayGov includes a range of payment modes such as

1. *Credit cards,*
2. *Debit cards,*
3. *Net Banking (Approximately 70 banks enabled),*
4. *Digital wallets,*
5. *Immediate Payment Service (IMPS),*
6. *Real-time gross payments (RTGS), and*
7. *National Electronics Funds Transfer System (NEFT).*

The PayGov payment involves few easy steps for making payments:

1. Applicant requires to login to the government portal compliant with PayGov platform,
2. The applicant enters the account or card details as demanded by the interface,
3. Further, funds are transferred to the nodal bank,
4. PayGov communicates the transactional details to the Government account,
5. Finally, the nodal bank transfers the funds to the government account, which marks the completion of the transactional process.

Furthermore, keeping in mind the data infrastructure and security concerns, the details are entered on a secured server such that no data theft or malicious activity could be performed. The data is coded before being transmitted over the internet to ensure no data is lost or tampered during the transactional activity.

6.3.8.2   Performance Evaluators (Fig. 6.11)

6.3.8.2.1   Increased Number of PayGov Users

PayGov has observed an upturn in terms of user participation. The platform has registered about 12,248,440 users till August 2016 leading to a revenue generation of INR 26,200,376,065 wherein INR 1,253,292,473 has been generated from the month of July itself. The data reveals the widespread use of PayGov due to the *convenience* level offered by it (Table 6.18).

Moreover, with the increased focus on *digitalization* and the intention to improve the overall *technology ecosystem* is leveraging the use of this platform. In addition, emphasis on the development of a financially sound economy has led to a huge implementation of the platform which in turn is expected to exhibit growth in the coming years as well.

Fig. 6.11   Performance evaluators of PayGov

Table 6.18   User Participation in terms of transactional amount generated and average cost incurred (data till 17 August 2016)

Duration	Number of users	Transactional amount (in Rupees)	Average cost incurred (in Rupees)
July 2016	499,327	1,253,292,473	2509.96
10–17 August 2016	183,346	242,103,626	1320.47
Till August 2016	12,248,440	26,200,376,065	2139.08
Till March 2017	17,693,320	51,590,227,780	

Source: http://paygovindia.gov.in/

Transactions (in thousands, 2016)

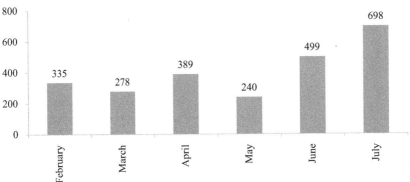

Fig. 6.12 Number of transactions in thousands during February–July 2016. (Source: Department of Electronics and Information Technology 2016)

6.3.8.2.2 More Number of PayGov Transactions

The number of online transactions has increased by leaps and bounds. The number of transactions made in February 2016 were 335,000 which increased to 698,000 in July 2016. This transition accounted for a 100 percent increase.

Therefore, an increase in the number of users due to rising penetration of different *internet-enabled* devices and *24×7 availability* has resulted in large number of transactions (Fig. 6.12).

6.3.8.2.3 Huge Revenue Generation

The platform registered increased number of users which led to large number of transactions, resulting in huge revenue generation. This upsurge signifies the acceptance towards the PayGov platform (Fig. 6.13).

The amount generated in six months has upscaled. In July 2016, the revenue encountered was INR 1200 million in comparison to INR 863 million in February 2016. Moreover, the platform showcases a positive future outlook, in correspondence to its present performance.

6.3.8.2.4 Its Alliances

A total number of 64 departments and agencies have integrated their online portal with the PayGov platform till December 2016. Further, integration with 86 departments has been initiated and with 28 departments is in the pipeline. Statistical data reveals that the integration of the 64

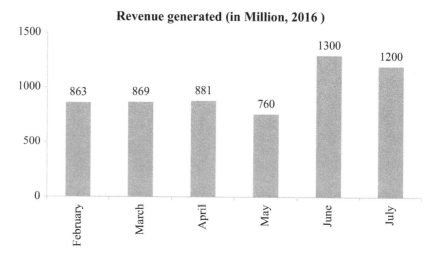

Fig. 6.13   Amount generated during February–July 2016. (Source: Department of Electronics and Information Technology 2016)

departments has generated revenue of INR 2620.04 crore which accounts for INR 40.93 crore on an average (Department of Electronics and Information Technology 2016) (Table 6.19).

Statistical data reveals that the integration of the 64 departments has generated revenue of INR 2620.04 crore (Table 6.20).

6.3.8.3 Case Summary

The platform has emerged as one of the largest aggregated platforms available in India. The platform helps to manage a wide range of day-to-day activities spanning from daily settlement activities to daily refund activities. It has many advantages which can help the citizens of the nation (Fig. 6.14).

1. *Daily settlement process*

   This process is carried out in all the government departments at the end of day to perform the required settlements. Furthermore, the incomplete transactions are also addressed.

2. *Daily payment process*

   The portal provides a 24×7 facility to make the payments. However, the entire amount is transferred to the government treasury/ bank account.

Table 6.19 State-wise departments compliant with PayGov (data accessed on 17 December 2016)

State	Department integrated
Bihar	Urban Development and Housing Department
Chhattisgarh	Chhattisgarh eProc
	State Portal
	CSIDC
Delhi	Delhi Jal Board (DJB)
	ILBS
	ILBS Conference
Haryana	Haryana CCTNS
Himachal Pradesh	Himachal State Portal
	Himachal E-District
	Himachal Pradesh Crop Insurance
	Shimla Municipal Corporation
Jharkhand	Commercial Tax
	Transport Tax
	Labor
	Ranchi Municipal Corporation
	Jharkhand Exam Board
	Jharkhand Registration
	Dhanbad Municipal Corporation
	Revenue, Registration and Land Reforms Dept
	Department of Mines and Geology, etc.
Karnataka	Nadakacheri—Karnataka
	M One
Kerala	Kerala CSC
	Kerala Forest
	Cochin Corporation
	Gov Kerala State Portal
	MG University
Lakshadweep	Lakshadweep Electricity
Madhya Pradesh	CPCT
	MP Online
Maharashtra	MAHAONLINE
	MCGM AQUA
	MCGM Property Tax
	MCGM Other Licenses, etc.
Manipur	MSPDCL Manipur
	Manipur Prepaid Online Electricity Recharge
Nagaland	Nagaland State Portal
Punjab	Punjab NRI
	State Portal
	Punjab Bureau Of Investment Promotion
Tamil Nadu	E-District
Uttar Pradesh	Uttar Pradesh CSC

Source: http://paygovindia.gov.in/merchants.html

Table 6.20 Different GOI departments and public sector undertaking compliant with the PayGov platform

GOI	CSC E-Governance Services India Limited
	National Internet Exchange
	NIELIT
	E-MSIPS
	Council of Architecture
	ERNET
	E-Basta
	E-Hospital-AIIMS
Public Sector Undertakings	Air India
	NTPC

Source: http://paygovindia.gov.in/merchants.html

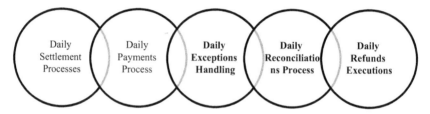

Fig. 6.14 PayGov-advantages. (Source: http://paygovindia.gov.in/about-pay-gov.html#operations-tab)

3. Daily exceptions handling
   This process occurs as and when needed. Thus, if any unusual problem is incurred, a special means to handle those problems is performed.
4. *Daily Reconciliations Process*
   This process ensures that no transaction is missed while compilation.
5. *Daily Refunds Executions*
   This process varies from department to department as some might offer an option to refund the fee and other might not.

6.3.9   Pradhan Mantri Jan-Dhan Yojana

6.3.9.1   Overview
In the past few years, the Indian economy has undergone numerous changes. The country has been categorized as the *newly industrialized country, one of the members of the G-20 (Group of Twenty)*, and *BRICS*

exhibiting an optimistic economic growth. Moreover, with the rise in the numerous government initiatives, one of them being Pradhan Mantri Jan Dhan Yojana (PMJDY), which has led to the augmentation in the overall financial functioning and infrastructure.

The honorable Prime Minister of India, Shri Narendra Modi, announced the introduction of the Pradhan Mantri Jan Dhan Yojana acting as a National Mission for Financial inclusion. The scheme was launched on 28 August 2014 by the Government of India. The name *Jan Dhan* was selected via an online competition held on the MyGov platform. Through this scheme, every citizen of the country can have a zero-balance account. Moreover, a maximum of two members in the family can have the zero-balance account. It revolves around availing oneself the facilities such as *Banking/Savings* and *Deposits account, remittance, pension, insurance, credit* conveniently and affordably. Furthermore, the account can be opened in any bank branch with zero balance, but if the account holder wishes to get a check book, he/she will have to satisfy the minimum account balance criteria. The scheme focuses on placing the citizens in a financially sound position, especially the poor masses (Table 6.21).

As in the words of our honorable Prime Minister Mr. Narendra Modi, "*Economic resources of the country should be utilized for the well-being of the poor. The change will commence from this point*" (pmjdy.gov.in 2014). This implies that the initiative is in the welfare of the society to enhance the overall status of the poor and low-income groups, to bring everyone at par. The journey so far entails the facts and figures shown in Table 6.22.

Swabhimaan Versus Pradhan Mantri Jan Dhan Yojana
Prior to the Pradhan Mantri Jan Dhan Yojana, an initiative called *Swabhimaan* was introduced focusing on the rural area only and had many loopholes per se *no technological mode* was incorporated, *tedious KYC formalities*, and so on. The PMJDY overcomes these issues and allows easy alignment of technology with the banking facilities leveraging the overall convenience.

Table 6.21 The Pradhan Mantri Jan Dhan Yojana status as on 28 December 2016

Accounts opened	26.20 Crore
Number of deposits made	Rs. 71,036.59 Crore
Suraksha Bima Policies	9.88 Crore
Jeevan Jyoti Bima Policies	3.08 Crore
Number of bank Mitras	1.26 lacs

Source: pmjdy.gov.in

Table 6.22 Distinguishing factors between the earlier approach and newer version

Swabhimaan (Earlier approach)	Pradhan Mantri Jan Dhan Yojana (Newer approach)
This scheme covered a limited geographical area, mostly spanned villages with a population of more than 2000.	It lays emphasis on the household along with Sub Service Areas (SSA), aiming at covering a larger area, in fact the whole country.
The target audience comprises of the rural area only.	The target audience covers both rural and the urban areas.
Bank Mitr/Bank Correspondents visited the villages on certain fixed days only.	This scheme showcased more active participation of the bank Mitr in terms of spreading awareness among the masses.
The scheme lacked interoperability of the accounts.	This scheme possesses the interoperability via the Rupay Debit cards and the Aadhaar-enabled Payment systems (AEPS).
Negligible focus on the technology.	Substantial focus on the technological aspect, accounts were primarily opened with the help of the Core Banking Solutions (CBS) available with the bank.
It does not entail any active involvement of the state and districts.	It encompasses the state and district level monitoring of the committees to be set up.
No focus on the financial literacy of the people.	It provides a separate Financial Literacy cell.

Source: https://www.pmjdy.gov.in/scheme

Slogan of the scheme

Mera khaata – Bhagya Vidhata.

My Bank Account – The Creator of the Good Fortune.

The scheme focuses on the long-term aspect to educate the backward section in context to the financial parameters and reduce the number of deceitful actions carried by the money-lenders and commission agents, thereby eradicating corruption.

Moreover, the accounts opened under this scheme are interlinked with the debit card provided under the RuPay scheme. RuPay scheme refers to the Indian Domestic card scheme, introduced by the National Payments Corporations of India (NPCI), offering *lower cost* and *affordability* coupled with *customized product offering* (National Payments Corporation of India 2016).

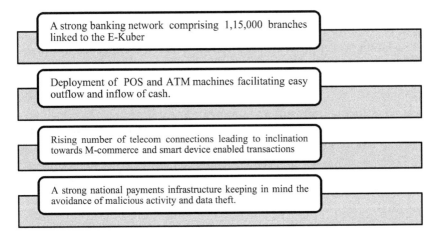

A strong banking network comprising 1,15,000 branches linked to the E-Kuber

Deployment of POS and ATM machines facilitating easy outflow and inflow of cash.

Rising number of telecom connections leading to inclination towards M-commerce and smart device enabled transactions

A strong national payments infrastructure keeping in mind the avoidance of malicious activity and data theft.

Fig. 6.15   Prerequisites for PMJDY. (Source: https://www.pmjdy.gov.in/scheme)

Prerequisites for transformational change in the digital financial inclusion (Fig. 6.15).

6.3.9.1.1   Objective
The objective of this plan is:

> *Ensuring access to various financial services like availability of basic savings bank account, access to need based credit, remittances facility, insurance and pension to the excluded sections i.e. weaker sections & low-income groups. This deep penetration at affordable cost is possible only with effective use of technology.* (pmjdy.gov.in 2016)

The objective emphasizes on the optimum utilization of the pool of resources such as the technological and financial means in the banking domain is to cope up with the changes taking place in the overall banking system.

6.3.9.1.2   Mission
Universal Access to Banking Facilities
The *linking* of approximately six lakh villages across the country and their districts with the Sub Service Area (SSA) to cater 1000–1500 households with the banking outlet. This is designed for the households to provide the accessibility to the different banking services within a reasonable distance of 5 km. This in turn will help to leverage the convenience levels and customer engagement.

Providing Basic Banking Accounts with Overdraft Facility and RuPay Debit Card to All Households
The account holder is provided with a RuPay Debit Card and an overdraft facility which is opened only after its satisfactory operation for a period of six months.

Financial Literacy Program
Financial literacy plays a crucial role for the users as an optimum utilization of the financial services is mandatory.

Creation of Credit Guarantee Fund
The scheme focuses on the formation of a credit guarantee account to help in covering the deficits and defaults made due to the overdraft accounts.

Micro-insurance
It safeguards the low-income people against specific risks and hazards involved in the exchange of money, up to 14 August 2017. Furthermore, continue this process on a continuous basis.

Unorganized Sector Pension Schemes like Swavalamban
This includes an individual should not be employed under the central and state government, autonomous body, or PSU of the Central and State governments. For instance, the Swavalamban Pension Yojana, which falls under the Pradhan Mantri Jan Dhan Yojana, is *a new pension scheme* being administered by the Interim Pension Fund Regulatory and Development Authority (PFRDA).

6.3.9.2   Performance Evaluators
The scheme has gained widespread acceptability by the people belonging to both the rural and the urban areas. Reasons for such an escalation can be credited to the rise in the standard of living of the people coupled with an increase in the awareness to safeguard themselves from misleading activities and secure their future in context to the financial aspects. Furthermore, the scheme has exhibited an increase in the number of the accounts opened both in the rural and the urban areas.

6.3.9.2.1   Increase in the Number of Accounts Opened
Indian financial system comprises of more than *27 Public Sector banks, 21 Private Sector banks* and *82 Regional Rural Banks,* which have been evolving

at a rapid pace. The introduction of the new improved technological infrastructure has led to an exponential growth in the smooth functioning of different transactional activities.

The performance review of this scheme reveals positive growth, as the number of accounts registered during the period January 2016 to May 2016 has shown an inclination. In addition, the scheme has enabled women, minorities, who were hitherto in a financially underprivileged situation, to enhance their overall financial stability (Table 6.23).

Furthermore, this upward trend is expected to continue in the coming years as well contributing to the rising concerns for social welfare and public expenditure (Fig. 6.16).

6.3.9.2.2   Trend of Zero Balance Account
The total depicts the grand total of the accounts opened in the various public sector, private sector and regional rural banks. With the trend of the zero balance accounts, it is quite evident that the awareness among the people regarding the importance of having bank accounts has exhibited an upsurge (Fig. 6.17).

Moreover, this implies a positive outcome in correspondence to the fulfillment of the objectives of this scheme. As the number of zero balance accounts depict that people are becoming a part of this financial inclusion. In addition, people are adding more to their knowledge about the numerous banking facilities they can avail themselves (Table 6.24).

6.3.9.2.3   Trend of RuPay Cards
National Payments Corporation of India introduced the Indian version of a debit or a credit card called the RuPay card. A RuPay card offers *easy affordability, low-cost transactions,* along with *interoperability* among the different payment channels and products (Table 6.25).

The number of RuPay card holders has witnessed an exponential growth which represents a healthy outlook for the scheme. The percentage change notifies the gradual change in the number of RuPay cards issued.

6.3.9.2.4   State-Wise Account Opening Report as on 21 December 2016 (Table 6.26)

6.3.9.3 Case Summary
Pradhan Mantri Jan Dhan Yojana is a major socio-economic initiative which marks a step towards a more financially sound Indian economy. Imparting the desired level of awareness in terms of different banking

Table 6.23 Accounts opened with Jan Dhan Yojana January–May 2016 (data as on 28 December 2016)

Bank Type	January		February		March		April		May	
	Rural	Urban	Rural	Urban	Rural	Urban	Rural	Urban	Rural	Urban
Public Sector Banks	89,537,898	71,298,083	92,238,015	73,014,075	94,292,464	74,165,273	95,192,409	75,238,730	96,345,344	75,988,933
Regional Rural Banks	31,248,470	5,153,529	31,932,869	5,278,875	32,571,057	5,357,511	32,916,141	5,408,110	33,362,156	5,480,446
Private Banks	4,478,240	2,962,470	4,537,976	3,004,730	4,843,513	3,045,656	4,924,815	3,072,318	5,014,006	3,101,415
Grand Total	125,264,608	79,414,082	128,708,860	81,297,680	131,707,034	82,568,440	133,033,365	83,719,158	134,721,506	84,570,794

Source: community.data.gov.in

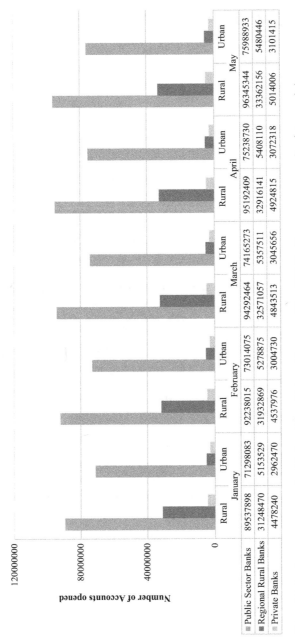

Fig. 6.16 Accounts opened with Jan Dhan Yojana (January–May 2016). (Source: community.data.gov.in)

	January		February		March		April		May	
	Rural	Urban	Rural	Urban	Rural	Urban	Rural	Urban	Rural	Urban
Public Sector Banks	89537898	71298083	92238015	73014075	94292464	74165273	95192409	75238730	96345344	75988933
Regional Rural Banks	31248470	5153529	31932869	5278875	32571057	5357511	32916141	5408110	33362156	5480446
Private Banks	4478240	2962470	4537976	3004730	4843513	3045656	4924815	3072318	5014006	3101415

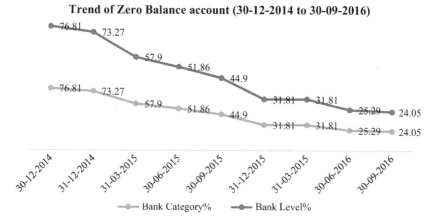

Fig. 6.17 Trend of the zero balance accounts opened with the Jan Dhan Yojana. (Source: https://www.pmjdy.gov.in/trend-zero)

services an individual can avail themselves with, the scheme is leading to a holistic development of the Indian society. In this knowledge era, with the innumerable banking services present, the plan emphasizes on the ideology of *Sab Ka Saath, Sab Ka Vikas.*

Further, financial inclusion helps in broadening the resource base of the financial system by inculcating a culture of savings among the rural population, hence resulting in an overall economic development of the country (Kaur & Singh 2015). Moreover, the scheme is expected to educate the individuals at the bottom of the pyramid in the society.

Consequently, the financial inclusion will enable the bank managers to comprehend the various needs of the lower section and participate in the government's poverty alleviation programs. Further, the scheme is focusing on the removal of middlemen and decreasing the leakage of funds helping to form a more secured and safer financial framework.

6.3.10   *DigiLocker/Digital Locker*

6.3.10.1 *Overview*
DigiLocker is simply "digital locker" service for the citizens launched by the Government of India in February 2015. It is one of the most important initiatives under the Digital India Program launched by the Government of India. The Department of Electronics and Information Technology (DeitY) is responsible for the launch and management of the initiative.

Table 6.24 Quarter-wise increase in the number of zero balance accounts from 30 September 2014 to 28 December 2016

Date	Number of zero balance accounts
30-09-2014	41,296,62
31-10-2014	52,173,633
29.11.2014	61,822,343
31.12.2014	**76,550,950**
Total	**138,373,293**
31-01-2015	84,473,729
28-02-2015	85,916,974
31-03-2015	85,213,334
29-04-2015	85,388,546
Total	**340,992,583**
Date	
27-05-2015	85,022,660
24-06-2015	85,176,376
22-07-2015	85,797,235
26-08-2015	80,376,439
Total	**336,372,710**
Date	
30-09-2015	74,771,502
28-10-2015	71,356,613
25-11-2015	67,670,775
30-12-2015	63,115,047
Total	**276,913,937**
Date	
27-01-2016	87,853,393
24-02-2016	60,642,439
30-03-2016	58,698,800
27-04-2016	57,200,764
Total	**264,395,396**
Date	
25-05-2016	56,271,712
29-06-2016	56,365,831
27-07-2016	54,758,680
31-08-2016	58,684,211
Total	**226,080,434**
Date	
28-09-2016	59,497,551
26-10-2016	59,336,401
30-11-2016	58,914,748
28-12-2016	63,216,560
Total	**240,965,260**

Source: https://www.pmjdy.gov.in/trend-zero

Table 6.25 Trend of Rupay cards

Date	31.12.2014	30.12.2015	28.12.2016	19.04.2017
No. of RuPay Debit Card	84,630,731	168,451,374	206,989,223	221,459,055
% Change	–	99.04%	23%	6.99%

Source: https://www.pmjdy.gov.in/Archive

Table 6.26 State-wise number of accounts opened with the Pradhan Mantri Jan Dhan Yojana as on 28 December 2016

S. No	State name	Rural accounts	Urban accounts	Deposit (in Crore)	Rupay Card issued
1	Andaman and Nicobar	35,546	16,064	23.65	42,548
2	Andhra Pradesh	4,565,392	4,175,328	1281.46	7,317,615
3	Arunachal Pradesh	126,295	81,844	65.27	167,636
4	Assam	8,829,545	2,915,300	2425.27	9,191,755
5	Bihar	18,798,527	10,226,020	5229.77	20,421,132
6	Chandigarh	33,435	182,143	102.7	181,210
7	Chhattisgarh	8,237,884	4,152,976	1774.19	8,428,157
8	Dadra and Nagar Haveli	68,149	10,552	25.31	48,519
9	Daman and Diu	17,924	12,680	10.41	19,779
10	Goa	106,312	39,161	93.68	121,961
11	Gujarat	5,510,688	5,093,507	2725.28	8,625,973
12	Haryana	3,255,251	2,737,180	2229.77	5,107,379
13	Himachal Pradesh	812,837	114,033	445.16	782,745
14	Jammu and Kashmir	1,752,752	405,935	312.23	1,678,056
15	Jharkhand	7,302,091	2,540,320	1721.49	7,339,656
16	Karnataka	6,259,189	4,388,777	2409.95	9,305,602
17	Kerala	1,524,295	1,710,765	1035.81	2,436,142
18	Lakshadweep	4413	380	5.29	3990
19	Madhya Pradesh	12,166,746	13,037,169	2916.32	18,265,459
20	Maharashtra	9,542,777	9,818,418	3763.35	14,191,781
21	Manipur	325,310	394,098	207.57	646,706
22	Meghalaya	325,445	66,253	209.02	214,849
23	Mizoram	107,164	175,208	36.81	98,104
24	Nagaland	105,902	100,291	37.77	168,712
25	Delhi	486,329	3,201,765	1438.21	2,985,641
26	Orissa	8,303,179	3,026,478	2808.81	8,863,288
27	Puducherry	66,339	74,298	33.46	124,227

(*continued*)

Table 6.26   (continued)

S. No	State name	Rural accounts	Urban accounts	Deposit (in Crore)	Rupay Card issued
28	Punjab	3,075,165	2,359,676	2284.89	4,618,713
29	Rajasthan	12,163,116	7,264,720	5083.45	16,638,002
30	Sikkim	63,483	21,442	26.93	67,077
31	Tamil Nadu	4,106,495	4,622,817	1441.61	7,345,140
32	Telangana	4,780,446	3,998,556	1269.31	7,441,043
33	Tripura	572,543	223,688	705.34	708,446
34	Uttar Pradesh	26,202,908	17,749,663	10,694.53	34,851,802
35	Uttarakhand	1,331,518	854,172	772.68	1,781,652
36	West Bengal	18,720,991	8,344,456	8313.41	21,228,558
	Grand Total	**169,686,381**	**114,136,133**	**63,960.16**	**221,459,055**

Sources: https://www.pmjdy.gov.in/statewise-statistics and pmjdy.gov.in

DigiLocker is a secure cloud-based platform that provides the opportunity to organizations and individuals to issue, verify and store their important legal documents. The purpose of a digital locker is to provide a personal electronic space to the public for storing their important documents such as residence documents.

The prerequisites of a DigiLocker are an Aadhaar card and a mobile number registered with it. The storage space of this account, which is 1 GB upgraded from maximum 10 MB at the time of launching, is linked to the Unique Identification Authority of India (Aadhaar number) of the user. Personal documents, such as university certificates, Permanent Account Number (PAN) cards, voter ID cards, can be stored in that space. This space can also be utilized to store the URIs of the E-Documents issued by various issuer departments. The documents can be saved in an XML format.

An extended facility associated with DigiLocker is E-Sign. The aim of the service is to reduce the use of physical documents and to improve the validity of the E-Documents. Government-issued documents can also be securely accessed with this service. The administrative expenses of government departments and agencies will be minimized, and services will be easily accessed by the residents. The user requires an Aadhaar card and a mobile number linked to it to create an account in DigiLocker.

The DigiLocker focuses on saving the documents online, thereby minimizing the usage of the physical documents. Further, it provides access to 1,65,36,64,482 authentic digital documents (DigiLocker 2017).

There are about 2.1 million registered users on the DigiLocker who have permanently saved their Permanent Account Number (PAN) cards, voter IDs and educational certificates.

6.3.10.2 Performance Evaluators
Within few years of its launch, DigiLocker has gained 4,676,392 users, 6,730,994 uploaded documents and 1,653,664,482 available documents. The E-Signed documents have risen to 369,264. There are 24 issuer organizations and 6 requestor organizations.

Table 6.27 depicts that the Aadhaar card is the most searched or issued document followed by the documents for registration of vehicles and LPG subscription voucher (Table 6.28).

Table 6.27 National Statistics for DigiLocker (data as on 10 March 2017)

Registered Users	4,676,392
Uploaded Documents	6,730,994
Available Documents	1,653,664,482
Issuer Organizations	24
Requestor Organizations	6
E-Signed Documents	369,264

Source: https://digilocker.gov.in/

Table 6.28 Document types available with DigiLocker (data as on 10 March 2017)

Document type	Available documents (issued or searchable)	Uploaded documents
Aadhaar Card	1,095,146,464	155,120
Registration of Vehicles	196,210,068	10,172
LPG Subscription Voucher	184,232,825	1477
Driving License	92,417,779	64,990
Income Certificate	29,454,629	3801
Caste Certificate	15,851,227	9113
Domicile Certificate	15,113,113	5525
Integrated Certificate	3,728,836	430
SSC Mark sheet—X	3,572,905	50,985
Community and Date of Birth Certificate	3,245,704	235
SSC Migration Certificate—X	2,871,076	313

Source: https://digilocker.gov.in/

Registrations

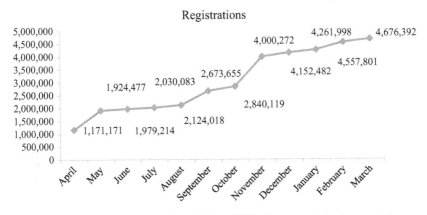

Fig. 6.18 Number of registrations 2016–2017. (Source: digilocker.gov.in)

6.3.10.2.1 Number of Registrations (2016–2017)

The number of registrations made in January 2016 was about 1,171,171, which have risen up to 4,261,998 in January 2017. This implies that the scheme has exhibited huge likeability by the citizens. Since its inception DigiLocker has witnessed an increase in the number of registrations. Thus, the coming years are expected to showcase a healthy growth owing to the numerous technological changes and an aim to bring at par the rural audiences with the urban (Fig. 6.18).

6.3.10.2.2 E-Signed Documents

The E-Signed documents help in preventing the misuse of data, by providing a safer and more secured means of data validation. Further, the number of E-Signed documents has exhibited an increase, which implies the acceptance of the initiative by the users.

Maharashtra leads the usage of this facility with 169,048 registrations followed by Uttar Pradesh, West Bengal, Andhra Pradesh, Tamil Nadu and the least registrations for DigiLocker are made in Mizoram and Lakshadweep (Fig. 6.19).

6.3.10.2.3 Decision Making

Under the Digital India program, feedbacks, comments and suggestions are solicited from the citizens for improvements in DigiLocker (MyGov 2016a, b). The suggestions and feedbacks of citizens about DigiLocker

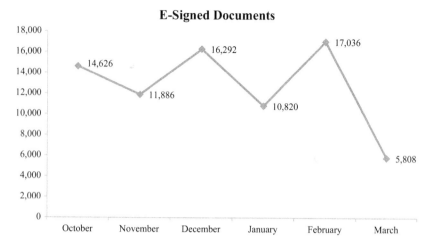

Fig. 6.19 Number of E-Signed documents 2016–2017. (Source: digilocker.gov.in)

are received on MyGov.in and the changes suggested are implemented by the government to make the service better. The citizens are empowered in the sense that they can anytime access their documents and can manage their documents online.

6.3.10.2.4 Digital Take Up

The DigiLocker facility under the Digital India program aims to provide shareable private space on a public cloud and allow digitization of documents. It stores all the documents of the citizens and makes them available when required. This facility helps in providing citizens their private space on a public cloud and this facilitates process re-engineering through paperless processes. Electronic format can be used to issue government documents and such government documents can be saved in E-Document government repositories. The facilities of 'E-Document repositories' and 'Digital Locker' help in improvement of citizen convenience and enabling paperless transactions in public services. Security is the prime motive for establishment of this service considering user authentication and audits. The citizens are benefitted by this service since they can easily access their documents through digital resources and also, they are not asked to present their documents in physical form in government offices/department/institution.

6.3.10.2.5   Digital Locker and Strategic Partnerships

DigiLocker Partners with CBSE
In 2016, CBSE launched an academic repository, *Parinam Manjusha*, to be used by employers and students to verify the academic records and certificates. The initiative is aligned to prevent the submission of fake documents. Further, documents contain a QR code, which can be verified with the help of mobile-based scanning application (*Hindustan Times* 2016).

DigiLocker Partners with Road Transport and Highways Ministry
In 2016, the Union Roads Ministry entered into a partnership with the IT ministry. The digital locker allows the users to leave their hard copies of the documents such as driving license and vehicle registration papers at home and save their soft copies online (LiveMint 2016a, b).

6.3.10.3 Case Summary
DigiLocker is a flagship program under Digital India initiative which aims at providing the citizens a shareable private space on a public cloud and making all documents/certificates available on this cloud. Further, the initiative makes it easier for the users to verify the authenticity of the registered users and allows an easy access to various documents.

Thus, the idea of allowing the easy submissions of documents has resulted in the increased usage of the platform, which is expected to showcase a growth in the coming years as well.

6.3.11   National Centre for Geo-Informatics (NCOG)

6.3.11.1 Overview
With the growing digitalization and an inclination towards the development of an intelligent transportation system, technological paradigm shift has been observed. Further, there has been a rise in the number of GPS-enabled automobiles to facilitate convenience in reaching the desired destination timely. Furthermore, there has been constant and conscious effort to introduce Geographic Information System into the various E-Governance and Mission Mode Projects. Thus, to support the growing concerns for an easy and smooth flow of traffic thereby reducing traffic congestion National Centre for Geo-Informatics was launched on 28 December 2015. NCOG works as an independent business division of

Media Labs Asia under the Department of Electronics and Information Technology (DeITY). In accordance to DeitY, the team will be led by T. P. Singh as Special Officer (from Bhaskaracharya Institute for Space Applications and Geo-Informatics [BiSAG], Gujarat) and Vinay Thakur as Director.

It provides a national platform to aid in the development of geo-informatics in the country. Furthermore, it allows the users to avail themselves the various geo-spatial applications and solutions at the local, state and central government levels. The various activities undertaken by the scheme are the development and maintenance of a unified, standardized and seamless Geographic Information System (GIS) database across the nation. It also focuses on the roll out of GIS-based governance processes to promote the transparency levels. The scheme has gained likeability and is expected to showcase a healthy outlook in the coming years owing to the numerous benefits it has to offer. In the words of the honorable minister of Electronics and Information Technology, Shri. Ravi Shankar Prasad,

> *GIS platform service is to be scaled up proactively for e-Governance at all levels under the Digital India Programme. (NCOG 2017)*

The project cost is INR 98.28 crores for a period of three years (*The Financial Express* 2017). The key projects associated with the scheme are Government Land Information System, Mining Surveillance System, Rural Electrification System, Saltpan Information System, PESA, National Directory, E-District CSCs, DI Outreach, DSS for AICTE, Road Information System, and so on.

6.3.11.2 *Performance Evaluators*

6.3.11.2.1   Current Status
It is the GIS which is prominently used by the various Central ministries, PSUs, PSEs to provide the location for the various land banks. By 2017, the mapping of the Ministry of Urban Development land banks for Delhi has been completed (Table 6.29).

6.3.11.2.2   Rural Electrification System (RES)
To locate and map the various villages without proper electrification in collaboration with the software developed and integrated called GARV (Grameen Vidyuti Karan) (Table 6.30).

Table 6.29   GLIS and its current status (data as on 10th March, 2017)

No. of lands parceled	35,000
No. of Ministries / Department	35
No. of Central Public Sector Enterprises	92
No. of officers trained	~1000

Source: ncog.gov.in

Table 6.30   RES and its current status (data as on 8 March 2017)

Villages Electrified	12,511
Villages to be electrified	5124
Unelectrified Villages	18,452

Source: garv.gov.in

6.3.11.2.3   PESA (Provision of the Panchayats Extension to the Scheduled Areas) Act 1996

The different states notified as the fifth schedule areas such as Andhra Pradesh, Chhattisgarh, Gujarat, Himachal Pradesh, Jharkhand, Madhya Pradesh, Maharashtra, Odisha, Rajasthan and Telangana have been mapped with PESA.

Further, the implementation of PESA enables the different fifth schedule areas to enhance their participation, reduce the poverty levels and minimize the exploitation of the tribal population (PESA 2017).

6.3.11.2.4   National Asset Directory (NAD)

It has been developed as a part of Panchayat Enterprise Suite under E-Panchayat Mission Mode Project. It acts as a repository and helps to keep a track of the different assets created, controlled, maintained by the Rural Local Bodies which include Panchayats at districts, intermediate and village level, Urban Local Bodies like Municipalities, corporations, town areas and Line department of the country. Further, it helps to assign a unique code to all the assets which in turn allow the effective utilization of the assets. By far all the assets have been identified and located on the map of India (The Asset Directory 2017).

6.3.11.2.5   Changes Incorporated in the Scheme

Earlier, the scheme had few data sets which were not compatible, however the changes amended in the scheme allow the compatibility of all the datasets at present. Thus, the scheme supports the different geo sets. Further, the authentication is also given utmost care which thereby prevents any malware of data.

6.3.11.2.6   Continuous Efforts to Facilitate a Smooth Understanding and Deployment of the Scheme

Workshop on the Use of GIS in Informed Planning and Decision Making
A consultation workshop was held from the 29 to 30 March 2016 at the India Habitat Centre, New Delhi, to discuss the implementation and requirements of the GIS system developed by the NCOG. The participants of the workshop comprised over 176 officials from the Centre and 34 states and UTs, academicians, NIC officials and State E-Mission Team together.

Workshop on Geo-Spatial Information System
The National Conference on Geo-Spatial Information System held on the 20 February 2015 with an aim to help in understanding the various GIS concepts, tools, and techniques. Further, to gain an exposure to the various digital image-processing techniques to perform the analysis of the geographical datasets.

6.3.11.3 Case Summary

NCOG promises a bright future and it has the potential to connect the citizens towards better services. Connecting the nation's geo-spatial data through this platform is an important activity and the response generated for NCOG has been really good from the citizen engagement point of view. It has been found to be really useful for the people from monitoring and status update point of view. This platform has been able to provide the services related to the geographical information systems in India and has been able to provide accurate information to various other applications.

6.3.12   National Scholarships Portal (NSP)

6.3.12.1 Overview

The National Scholarship Portal is a Mission Mode Project under the National E-Governance Plan (NeGP). The portal is a one-stop solution that provides access to a large number of services spanning from the filling of the student application form to the distribution of the scholarships.

The portal is centered on the provision of Automation, Streamlining and Effective Management of the numerous processes associated with the application receipt, processing, sanction and disbursal of centrally sponsored scholarship schemes to students.

The platform is aligned towards a SMART approach, which implies the provision of a Simplified, Mission-oriented, Accountable, Responsive and Transparent delivery of the services available. Keeping in mind that the scheme primarily focuses on the students, a simple registration process has been implemented from the academic session 2016–2017 wherein the availability of Aadhaar cards has been made optional. Thus, registrations can be done easily with the help of either their enrollment numbers or bank pass book.

6.3.12.2 Performance Evaluators

6.3.12.2.1   Current Status for National Scholarship Portal
The scheme provides the facility to check the All India Survey for Higher Education Code (AISHE), which covers about 1,617,084 registered Universities/Boards/Institutions and allows the students to apply for renewal of scholarships. Till March 2017, total 13,311,583 fresh registrations have been made and 6,276,713 renewals have been applied for (Table 6.31).

6.3.12.3 Case Summary
The National Scholarship Portal is an integrated portal which provides simplified services to all the students and helps in improving the transparency of data and the standardization process. The system suggests the scholarships available for the students, scholarship processing and create master data for institutions ad courses at all India level.

The scheme also serves as a Decision Support System for the different Ministries and Departments. It also facilitates monitoring of the scholarship disbursement, that is, from student registration to the delivery of

Table 6.31   Current status for National Scholarships Portal

Ministries	16
Departments/Councils Registered	23
Schemes	
Registered Universities/Boards/Institutions	1,617,084
Total Registrations	Fresh: 13,311,583 Renewal: 6,276,713

Source: http://scholarships.gov.in/

funds with the help of a comprehensive MIS. It also offers a complaint redressal portal wherein the students can share their grievances and get them resolved by the authorities. Furthermore, the scheme is estimated to showcase an optimistic growth in the future as well owing to the large number of features delivered by it.

6.4   CONCLUSION

India is at the cusp of a massive digital revolution. With the emerging digital technology and computer networks arises the need to cope up with the upcoming technological changes and advancements. Digital India encompasses a diverse range of schemes and platforms with a vision to digitally transform and empower the society using IT as a growth engine.

It is evident that India and its villages are moving in a direction wherein there is a slow yet steady connection among the rural and the urban areas. In addition, the initiatives emphasize on bridging the technological gap that exists among the rural and the urban areas, which would result in a robust knowledge economy.

The myriad of initiatives undertaken by the Government of India emphasize on the provision of the digitization services for the scanned document images or physical documents for any organization, reduce data redundancy and promote paperless transactions along with cooperation and suggestions from its citizens. Further, the initiatives aim at increasing the network penetration ultimately filling the connectivity gaps that prevail across the nation. The idea behind the creation of a digital infrastructure is to move beyond the passive levels of information leading to a more active citizen participation in the decision-making process.

APPENDIX: VARIOUS E-GOVERNANCE INITIATIVES TAKEN IN DIFFERENT STATES AND UNION TERRITORIES OF INDIA

State/UTs	Initiatives
Andhra Pradesh	MeeSeva
	Complaint Redressal System
	Employee Information System (EIM)-Department of School Education
	Prajavani (AP)—An E-Effort to Empower
Arunachal Pradesh	Community Information Centers (CICs)
	Online Bus schedule services
Assam	Electoral Rolls
	Public Utility Forms
	Passport Computerization System
	National Register of Citizenship (NRC)
	Prithvi Geographical Information System (GIS)
	E-UdyogRatna
	VIDHAN Magistracy Case Management System
	JANA-SEWA Computerized Public Services Facilitation Counter System
	PARISHODH Computerized Bakijai (Loan Repayment) System
	NATHI-AWASTHITI Computerized File Monitoring System
	ANUSHRAWAN Computerized Circle Office to DC Office Monthly Reporting System
	MANAB-SAMPAD Computerized Personnel Information Management System
	GRIHA-LAKSHMI Computerized Public Distribution System (PDS)
	BATON Computerized Payroll System
	ABHIYOG Computerized Public Grievance System
	GRAMUNNAYAN Computerized DRDA Schemes Monitoring System
	DAK Mail Management Application

(continued)

© The Author(s) 2019
S. K. Muttoo et al., *E-Governance in India*,
https://doi.org/10.1007/978-981-13-8852-1

(continued)

State/UTs	Initiatives
Bihar	Jankari
	E-Gazette
	Online Electricity Bill Payment
	Online Enrollment in Electoral Roll
	Information and Public Relations Department
Chandigarh	E-Jan Sampark
	E-Sampark
	Gram Sampark—Rural Knowledge Centre
	m-Sampark
Chhattisgarh	Online Lands Records
	Online Grievance Redressal
	Online Electoral Rolls
	Choice—Chhattisgarh Online Information for Citizen Empowerment
	E-Mail Directory
	E-Challan
Delhi	Grievance Redressal
	Application status finder
Goa	Mutation as part of DHARANI—Land Records Information System
Gujarat	Mahiti Shakti
	Online Application Forms
	Government Resolutions (GR) Book Online
	Gujarat Bank of Wisdom
	E-City—A one-stop civic shop
	Jan Seva Kendra
	E-Dhara
	SWAGAT—State-wide Attention on Public Grievances by Application of Technology
	E-Gram Viswa Gram Project
Haryana	Online Land Records—BhuLekh/Jamabandi
Himachal Pradesh	E-Samadhan—Online Public Grievance Solution
	Sugam—Integrated Community Information Centre
	Write to Chief Minister
	E-Gazette
	HP Police Web portal
	E-Salary
Jammu & Kashmir	Community Information Center—CIC
	Online Employment Exchange Information
	Online Motor Vehicle Information
	Information related to the Forms and Procedures
Jharkhand	E-Rahat Emergency Corner
	Grievance Redressal—Samvad aur Samadhan
	E-Nibandhan

(continued)

(continued)

State/UTs	Initiatives
Karnataka	Online Land Records E-NagrikSeva Common Service Centre Gyanshila Bhoomi SamanyaMahiti E-Granthalaya SahakaraDarpana of Directorate of Co-operative Audit Sarathi&Vahan of Transport Department Returns Filing System (RFS) E-MAN AasthiTerige (Property Tax) Child Labour Eradication Activities Information System (CLEAIS) KrishiMarataVahini Ahara Audit Monitoring System (AMS) RaitaMitra CASCET—2003 E-Archive
Kerala	Akshaya BhuRekha FRIENDS
Madhya Pradesh	Online Grievance Redressal Tele Samadhan Child Record Information System Citizen Charters
Maharashtra	Gyandoot RojgarWahini SARITA—Stamps & Registration Information Technology based Administration SETU—Integrated Citizen Facilitation Centres Kalyan-Dombivali Municipal Corporation (KDMC)-Citizen Facilitation Centers (CFC)
Manipur	Health Services Online Government Notification
Mizoram	Mizoram Gazette
Meghalaya	Online VAT Application Online agriculture market price Online Public Utility Forms Online constituency-wise electoral roll
Nagaland	Online Government Circulars and notifications Online Public Utility Forms Online Voter List

(continued)

(continued)

State/UTs	Initiatives
Orissa	Bhulekh—Land Record Web portal of Odisha
	E-Shishu
	Download Forms Website
	ITIMS—Integrated Transport Information Management System
	ORIS— (Odisha Registration Information System)
	E-Gram—Rural Information Gateway
	E-Literacy
Punjab	Punjab Sewa Online
	Suwidha
	Online Public Utility Forms
	Land Records Management System
	Saarthi and vahan
	E-District
Rajasthan	E-Mitra
	Raj stamps
	Send your queries to Chief Minister
Sikkim	Online Public Utility Forms
	Online Voters list
Tamil Nadu	Online Land Records
	Employment Online
Tripura	Hospital Management System
	Agartala Municipal Corporation
	Public Utility Forms
	E-Suvidha—Service Facilitation Centre (SFC)
Uttar Pradesh	E-Scholarship
	Bhulekh, UP
	Koshwani
	Court case Information System
	GIS based Planning Atlas
	Lokvani
Uttarakhand	Devbhoomi—Uttarakhand Land Records
	Online Content Creation/IT Enabled Course Curriculum
	School Education Portal
West Bengal	Telemedicine: Midnapore
	Smart Card
	Computerization of Government Departments
	GIS for Municipalities
	Public interfaces through Info kiosks/websites
	Higher Education Department
	Tourism Department
	Information & Cultural Affairs Department
	Geographical Information System
	Connectivity
	West Bengal State-wide Area Network (WBSWAN)
	IT Enabled Braille Education for the Blind Schools of West Bengal and Augmentation of Infrastructure
	Infrastructure Development

BIBLIOGRAPHY

Alam Siddiquee, N. (2006). Public Management Reform in Malaysia: Recent Initiatives and Experiences. *International Journal of Public Sector Management, 19*(4), 339–358.

Alawadhi, N. (2016). *MyGov Set to Implement Big Data, Analytics*. Gadgets Now. Available at http://www.gadgetsnow.com/tech-news/MyGov-set-to-implement-Big-Data-analytics/articleshow/45545698.cms. Accessed 26 Nov 2016.

Alawneh, A., Al-Refai, H., & Batiha, K. (2013). Measuring User Satisfaction from e-Government Services: Lessons from Jordan. *Government Information Quarterly, 30*(3), 277–288.

Ali, H. E. (2015). Special Issue on e-Government in Developing and Transitional Countries. *International Journal of Public Administration in the Digital Age, 2*(3), iv–iv.

Al-Qirim, N. A., & Corbitt, B. J. (2004). The Government and e-Governance: A Policy Perspective on Small Businesses in New Zealand. In *e-Business, e-Government & Small and Medium-Size Enterprises: Opportunities and Challenges* (pp. 1–18). Hershey: IGI Global.

Amagoh, F. (2015). An Assessment of e-Government in a West African Country: The Case of Nigeria. *International Journal of Public Administration in the Digital Age (IJPADA), 2*(3), 80–99.

Anand, R., Medhavi, S., Soni, V., Malhotra, C., & Banwet, D. K. (2018). Transforming Information Security Governance in India (A SAP-LAP Based Case Study of Security, IT Policy and e-Governance). *Information & Computer Security, 26*(1), 58–90.

Attendance.gov.in. (2017). Retrieved from http://attendance.gov.in. Accessed 15 Mar 2017.

© The Author(s) 2019
S. K. Muttoo et al., *E-Governance in India*,
https://doi.org/10.1007/978-981-13-8852-1

181

Bannister, F., & Walsh, N. (2002). The Virtual Public Servant: Ireland's Public Services Broker. *Information Polity, 7*(2, 3), 115–127.

Bouckaert, G. (2006). Modernising Government: The Way Forward—A Comment. *International Review of Administrative Sciences, 72*(3), 327–332.

Bovaird, T. (2005). Performance Measurement and Evaluation of e-Government and e-Governance Programmes and Initiatives. In *Practicing e-Government: A Global Perspective* (pp. 16–42). Hershey: IGI Global.

Cabinet Office. (2007). *e-Government Unit*. e-Government. Available at http://webarchive.nationalarchives.gov.uk/20100807034701/ http://archive.cabinetoffice.gov.uk/e-government/. Accessed 11 July 2016.

Chakrabarty, T. (2008). *Towards an Ideal e-Governance Scenario in India*. TCS. Available at http://www.tcs.com/SiteCollectionDocuments/White%20Papers/tcs_government_idealegovernanceindia.pdf. Accessed 10 July 2016.

CIFPA. (2013). *Good Governance in the Public Sector—Consultation Draft for an International Framework*. Available at http://www.ifac.org/system/files/publications/files/Good-Governance-in-the-Public-Sector.pdf. Accessed 11 July 2016.

Community.data.gov.in. (2016). *Accounts Opened Under Pradhan Mantri Jan-Dhan Yojana (PMJDY) Upto 25.05.2016*. https://community.data.gov.in/accounts-opened-under-pradhan-mantri-jan-dhan-yojana-pmjdy-upto-25-05-2016/. Accessed 27 Dec 2016.

Datta, S. K., & Singh, K. (2018). Aspects of Inclusion and Peoples' Empowerment Related to National Rural Employment Guarantee Scheme in India. *International Journal of Public Administration, 41*(2), 83–94.

DeitY. (2016a). *National e-Governance Division*. e-Governance. Available at http://deity.gov.in/content/national-e-governance-division. Accessed 10 July 2016.

DeitY. (2016b). *National e-Governance Plan*. e-Governance. Available at http://deity.gov.in/content/national-e-governance-plan. Accessed 11 July 2016.

DeitY. (2016c). *Institutional Mechanism*. Standards & Policies, Projects & Initiatives. Available at http://deity.gov.in/content/institutional-mechanism. Accessed 12 July 2016.

Deloitte. (2016). *e-Governance and Digital India Empowering Indian Citizens Through Technology*. ASSOCHAM. Available at http://www2.deloitte.com/content/dam/Deloitte/in/Documents/technology-media-telecommunications/in-tmt-empowering-indian-citizens-through-technology-noexp.pdf. Accessed 15 Nov 2016.

Department of Electronics & Information Technology. (2016). *PayGov*. Retrieved from http://paygovindia.gov.in/about-paygov.html#process-tab. Accessed 19 Dec 2016.

Dhamija, A., & Dhamija, D. (2018). Motivational Factors and Barriers for Technology-Driven Governance: An Indian Perspective. In *Innovative*

Perspectives on Public Administration in the Digital Age (pp. 194–211). Hershey: IGI Global.

DigiLocker. (2017). Retrieved from https://digilocker.gov.in/public/dashboard. Accessed 14 Mar 2017.

Digital India. (2016a). *2 Years of Achievement*. Department of Electronics and Information Technology. Available at http://digitalindia.gov.in/ebook/deity/page4.php. Accessed 26 Nov 2016.

Digital India. (2016b). *e-Governance Policy Initiatives Under Digital India*. What's New? Available at http://www.digitalindia.gov.in/content/e-governance-policy-initiatives-under-digital-india. Accessed 10 July 2016.

Digital India. (2016c). *Information for All*. Programme Pillars. Available at http://www.digitalindia.gov.in/content/information-all. Accessed 15 Nov 2016.

DIT & NCAER. (2006). *INDIA: e-Readiness Assessment Report 2006*. Ministry of Communications and Information Technology, Government of India. Available at https://negp.gov.in/pdfs/e-Rediness_Report_2006.pdf. Accessed 21 July 2016.

DIT & NCAER. (2007). *INDIA: e-Readiness Assessment Report 2005*. Ministry of Communications and Information Technology, Government of India. Available at https://negp.gov.in/pdfs/e-Readiness_Report_2005.pdf. Accessed 21 July 2016.

DIT & NCAER. (2010). *INDIA: e-Readiness Assessment Report 2008*. Ministry of Communications and Information Technology, Government of India. Available at http://www.doitc.rajasthan.gov.in/administrator/Lists/Downloads/Attachments/19/e-Readiness_20Report_202008.pdf. Accessed 15 Nov 2016.

Dutta & Devi. (2016). *e-Governance Status in India*. Researchgate. Available at https://www.researchgate.net/profile/Ajay_Dutta/publication/2833 19107_e-Governance_Status_in_India/links/563304d308ae242468 da08d1/e-Governance-Status-in-India.pdf. Accessed 4 Oct 2016.

Dutta, D., & Dutta, D. (2018). *Development Under Dualism and Digital Divide in Twenty-First Century India*. Singapore: Springer.

e-Hospital. (2017). Retrieved from http://ehospital.nic.in/ehospitalcloud/CloudHome/faq.jsp. Accessed 17 Mar 2017.

etaal.gov.in. Available on http://etaal.gov.in/etaal/MMPIndex.aspx. Retrieved 18 Feb 2017.

Federal bank. (2017). Retrieved from http://www.federalbank.co.in/selfie. Accessed 10 Mar 2017.

Ganore, P. (2011). *e-Governance – The Way to the Efficiency and Transparency in India*. ESDS. Available at http://www.esds.co.in/blog/e-governance-electronic-government-the-way-to-the-efficiency-and-transparency-in-india/#sthash.FLf025Cw.dpbs. Accessed 11 July 2016.

Ghosh. A. (2016). *The Future for e-Governance – Mobile First, Citizen Centric and Smart!*. e-Governance. Available at http://blogs.nasscom.in/the-future-for-e-governance-mobile-first-citizen-centric-and-smart/. Accessed 15 Nov 2016.

Giap, T. K., Randong, Y., & Mu, Y. (2013). *Annual Analysis of Competitiveness, Simulation Studies and Development Perspective for 34 Greater China Economies: 2000–2010*. Singapore: World Scientific.

Gidwani, K. (2016). e-Governance: Changing Scenario of Public Services. In *Trends, Prospects, and Challenges in Asian e-Governance* (pp. 192–207). Hershey: IGI Global.

Gisselquist, R. (2012). *What Does Good Governance Mean*. United Nations University. Available at http://unu.edu/publications/articles/what-does-good-governance-mean.html. Accessed 11 July 2016.

Graham, J., Amos, B., & Plumptre, T. (2003). Principles for Good Governance in the 21st Century. *Policy Brief, 15*(2003), 1–6.

GrameenVidyutikaran. (2017). Retrieved from http://garv.gov.in/dashboard/ue. Accessed 9 Mar 2017.

Gulati, A. G. (2006). Singapore-An e-Government Delight and Managing the Same for India. *Indian Journal of Public Administration, 52*(2), 202–221.

Gupta, R. (2014). Study and Proposal of Wearable Computer Applications. *Research Journal of Science and IT Management, 3*(4), 87–95.

Gupta, I. C., & Jaroliya, D. (2008). *It Enabled Practices and Emerging Management Paradigms* (p. 195). New Delhi: Excel Books.

Gupta, R., & Muttoo, S. K. (2016). Internet Traffic Surveillance & Network Monitoring in India: Case Study of NETRA. *Network Protocols and Algorithms, 8*(4), 1–28. https://doi.org/10.5296/npa.v8i4.10179.

Gupta, R. & Pal, S. K. (2016a, December). Data Analytics for improvement of Click-through Rate for Digital India Campaign's Effectiveness. In *Young Scientist Conclave, India International Science Festival (IISF)-2016*. Ministry of Science & Technology, Vijnana Bharati, ICAR, & CSIR-National Physics Lab.

Gupta, R., & Pal, S. K. (2016b, December). Trans-jurisdictional Water Pollution Management in India. In *Young Scientist Conclave, India International Science Festival (IISF)-2016*. Ministry of Science & Technology, Vijnana Bharati, ICAR, & CSIR-National Physics Lab.

Gupta, R., & Pal, S. K. (2017a). A Study on Cause Related Marketing: Antecedents and Consequents in Relation to Purchase Intention of Consumers in Delhi, India. *Ansal University Business Review (AUBR), 5*(1), 132–164. ISSN 2320-0502.

Gupta, R., & Pal, S. K. (2017b). Converting Apprehensive Customers to Willing Customers: Building Trust in Online Shopping Arena. *IIMS Journal of Management Science, 8*(2), 198–220.

Gupta, R., & Pal, S. K. (2019). Click-Through Rate Estimation Using CHAID Classification Tree Model. In *Advances in Analytics and Applications* (pp. 45–58). Singapore: Springer.

Gupta, R., & Pathak, C. (2014). A Machine Learning Framework for Predicting Purchase by Online Customers based on Dynamic Pricing. *Procedia Computer Science, 36*, 599–605. Elsevier.

Gupta, R., Aggarwal, A., & Pal, S. K. (2012). Design and Analysis of New Shuffle Encryption Schemes for Multimedia. *Defence Science Journal, 62*(3), 159–166. https://doi.org/10.14429/dsj.62.1008.

Gupta, R., Muttoo, S. K., & Pal, S. K. (2013a). Analysis of Information Systems Security for e-Governance in India. In *National Workshop on Cryptology-2013* (pp. TSII 17–25). DESIDOC, DRDO & CRSI.

Gupta, R., Muttoo, S. K., & Pal, S. K. (2013b). Awareness, Adoption and Acceptance of e-Government Services in India. In *International Conference on Research in Marketing (ICRM 2013)* (pp. PS2/1–11). IIT Delhi, XLRI & Curtin University.

Gupta, R., Chaudhary, N., & Pal, S. K. (2014a, September). Hybrid Model to Improve BAT Algorithm Performance. In *Advances in Computing, Communications and Informatics (ICACCI, 2014)* (pp. 1967–1970). New York: IEEE Xplore.

Gupta, R., Chaudhary, N., Garg, A., & Pal, S. K. (2014b). Efficient and Secured Video Encryption Scheme for Lightweight Devices. *INROADS- An International Journal of Jaipur National University, 3*(1s2), 346–351. http://www.indianjournals.com/ijor.aspx?target=ijor:inroads&volume=3&issue=1s2&article=015.

Gupta, R., Muttoo, S. K., & Pal, S. K. (2014c, September). Proposal for Integrated System Architecture in Utilities. In *Advances in Computing, Communications and Informatics (ICACCI, 2014)* (pp. 1995–1998). New York: IEEE Xplore.

Gupta, R., Pal, S. K., & Chaudhary, N. (2014d). A Novel Lightweight Encryption Scheme for Multimedia Data. *International Journal of Enhanced Research in Science Technology & Engineering, 3*(1), 91–96.

Gupta, R., Muttoo, S. K., & Pal, S. K. (2015a). Analyzing Security Checkpoints for an integrated Utility based Information System. *2015 Emerging Research in Computing, Information, Communication and Applications (ERCICA), 3*(1), 350–355. Singapore: Springer.

Gupta, R., Muttoo, S. K., & Pal, S. K. (2015b). Multimedia Data Challenges for Digital India Campaign. *CSI Communications, 39*(6), 19–21. [Computer Society of India].

Gupta, R., Muttoo, S. K., & Pal, S. K. (2015c). Review Based Security Framework for e-Governance Services. *Chakravyuh, 11*(1), 42–50. [SAG Lab, DRDO].

Gupta, R., Muttoo, S. K., & Pal, S. K. (2015d). Understanding Data Mining Applications for e-Governance. *CSI Communications, 39*(9), 14–17. [Computer Society of India].

Gupta, R., Muttoo, S. K., & Pal, S. K. (2015e, December). Dynamic Route Map Generation Scheme for Mobiles. In *2015 12th Annual IEEE India Conference (INDICON)* (pp. 1–6). Delhi, India: IEEE Xplore.

Gupta, R., Muttoo, S. K., & Pal, S. K. (2016a). Application of Web Mining & Analytics for Improving e-Governance in India. In *Web Usage Mining Techniques and Applications Across Industries* (pp. 223–247). IGI Global. https://doi. org/10.4018/978-1-5225-0613-3.ch009.

Gupta, R., Muttoo, S. K., & Pal, S. K. (2016b). BAT Algorithm for Improving Fuzzy C-Means Clustering for Location Allocation of Rural Kiosks in Developing Countries Under e-Governance. *Egyptian Computer Society Journal, 40*(2), 77–86.

Gupta, R., Muttoo, S. K., & Pal, S. K. (2016c). Binary Division Fuzzy C-Means Clustering and Particle Swarm Optimization based Efficient Intrusion Detection for e-Governance Systems. *International Review on Computers and Software (IRECOS), 11*(8), 672–681. https://doi.org/10.15866/irecos. v11i8.9546.

Gupta, R., Muttoo, S. K., & Pal, S. K. (2016d). *Data Analytics for 'Digital India' Branding for Improved e-Governance Services* (Vol. 1(1), pp. 83–84). Conference on Brand Management 2016, IIT Delhi, Emerald India.

Gupta, R., Muttoo, S. K., & Pal, S. K. (2016e). *e-Governance in Emerging Economy*. Scholars World, Daya Publishing House (Astral International Pvt. Ltd.), Thomson Press India Limited. ISBN: 978-93-5130-979-6. http:// astralint.com/bookdetails.aspx?isbn=9789351309796.

Gupta, R., Muttoo, S. K., & Pal, S. K. (2016f). Implementation & Analysis of Integrated Utility System in Developing Nation Like India. *International Journal of Recent Contributions from Engineering, Science & IT (iJES), 4*(2), 11–17. https://doi.org/10.3991/ijes.v4i2.5757.

Gupta, R., Muttoo, S. K., & Pal, S. K. (2016g). *Understanding Fraudulent Activities Through M-Commerce Transactions* (pp. 68–96). IGI Global: Securing Transactions and Payment Systems for M-Commerce. https://doi. org/10.4018/978-1-5225-0236-4.ch004.

Gupta, R., Muttoo, S. K., & Pal, S. K. (2016h, December). e-Participation Assessment in Indian States to Track e-Governance Development. In *Young Scientist Conclave, India International Science Festival (IISF)-2016*. Ministry of Science & Technology, Vijnana Bharati, ICAR, & CSIR-National Physics Lab.

Gupta, R., Pal, S. K., & Muttoo, S. K. (2016i). Design & Analysis of Clustering Based Intrusion Detection Schemes for e-Governance. In *The International Symposium on Intelligent Systems Technologies and Applications* (pp. 461–471). Cham: Springer.

Gupta, R., Pal, S. K., & Muttoo, S. K. (2016j). Network Monitoring and Internet Traffic Surveillance System: Issues and Challenges in India. In *Intelligent Systems Technologies and Applications* (pp. 57–65). Cham: Springer.

Gupta, R., Muttoo, S. K., & Pal, S. K. (2017a). Development of e-Governance in Emerging Economy Like India: Assessment and Way Ahead for Key Components. In *10th International Conference on Theory and Practice of*

Electronic Governance (ICEGOV2017) (pp. 1–4). ACM, United Nations University & NEGD GOI.

Gupta, R., Muttoo, S. K., & Pal, S. K. (2017b). Fuzzy C-Means Clustering and Particle Swarm Optimization-based Scheme for Common Service Center Location Allocation. *Applied Intelligence, 47*(3), 624–643. https://doi.org/10.1007/s10489-017-0917-0 SPRINGER.

Gupta, R., Muttoo, S. K., & Pal, S. K. (2017c). Proposed Framework for Information Systems Security for e-Governance in Developing Nations. In *10th International Conference on Theory and Practice of Electronic Governance (ICEGOV2017)* (pp. 1–2). ACM, United Nations University & NEGD GOI.

Gupta, R., Muttoo, S. K., & Pal, S. K. (2017d). *Techniques for Improvement of e-Governance in Developing Nations.* Doctoral Thesis, University of Delhi.

Gupta, R., Muttoo, S. K., & Pal, S. K. (2017e). The Need of a Development Assessment Index for e-Governance in India. In *10th International Conference on Theory and Practice of Electronic Governance (ICEGOV2017)* (pp. 1–9). ACM, United Nations University & NEGD GOI.

Gupta, R., Pal, S. K., & Muttoo, S. K. (2017f). Citizen Relationship Management by the Government of India Through Social Media Channels. In *Contemporary Issues in Social Media Marketing* (pp. 1–30). Routledge. https://www.routledge.com/Contemporary-Issues-in-Social-Media-Marketing/Rishi-Bandyopadhyay/p/book/9781138679184.

Gupta, R., Muttoo, S. K., & Pal, S. K. (2018a). A Comprehensive Review of Approaches Used for Solving Tele-center Location Allocation Problem in Geographical Plane. *International Journal of Operational Research*, Accepted but yet to be published, INDERSCIENCE.

Gupta, R., Yadav, S., Prasad, A., & Pal, S. K. (2018b). Measuring Data Quality across Open Government Datasets. *Facets of Business Excellence (FOBE) in IT, 1*(1), 442–451.

Harpreet, K., & Kawal, S. (2015). *Pradhan Mantri Jan Dhan Yojana (PMJDY): A Leap Towards Financial Inclusion in India.* http://www.ermt.net/docs/papers/Volume_4/1_January2015/V4N1-108.pdf. Accessed 27 Dec 2016.

Hindustan Times. (2016). *CBSE starts Online Repository for Students to View Mark Sheets.* Available at https://www.hindustantimes.com/mumbai-news/cbse-starts-online-repository-for-students-to-view-mark-sheets/story-YZxvP2x3sk-PjRrDcK5rMeO.html. Accessed 21 Jan 2017.

Hindustan Times. (2017). *DigiLocker Partners with CBSE.* Available at https://www.hindustantimes.com/mumbai-news/cbse-starts-online-repository-for-students-to-view-mark-sheets/story-YZxvP2x3skPjRrDcK5rMeO.html. Accessed 15 Jan 2017.

Hirwade, M. A. (2010). Responding to Information Needs of the Citizens Through e-Government Portals and Online Services in India. *The International Information & Library Review, 42*(3), 154–163.

Hirwade, M. A. (2012). Implementation of e-Governance in India. In *Encyclopedia of Cyber Behavior* (pp. 1282–1304). Hershey: IGI Global.

Howard, M. (2001). e-Government Across the Globe: How Will'e'change Government. *e-Government, 90*, 80.

http://www.digitalindia.gov.in. (2017). Retrieved from http://www.digitalindia.gov.in/content/broadband-highways. Accessed 14 Mar 2017.

Husain, M. S., & Khanum, M. A. (2017). Cloud Computing in e-Governance: Indian Perspective. In *Securing Government Information and Data in Developing Countries* (pp. 104–114). Hershey: IGI Global.

India.gov.in. (2015). *National e-Governance Plan.* Available at https://india.gov.in/e-governance/initiatives/central-initiatives. Accessed 10 July 2016.

Invest India. (2017). *IT & ITeS.* Retrieved from http://www.investindia.gov.in/it-and-ites-sector/. Accessed 22 Mar 2017.

Jain, P. K. (2016). Reformation in Mining Sector: A National Perspective. *Mineral Economics, 29*(2–3), 87–96.

Jain Gupta, P., & Suri, P. (2017). Measuring Public Value of e-Governance Projects in India: Citizens' Perspective. *Transforming Government: People, Process and Policy, 11*(2), 236–261.

Jeevan Pramann. (2017). *Key Performance Indicators.* Retrieved from https://jeevanpramaan.gov.in/signin. Accessed 10 Apr 2017.

Joshi, D., & Bansal, T. (2017, July). M Learning Apps for Digital India. In *Computing Conference, 2017* (pp. 1136–1142). London, UK: IEEE.

Jun, M. (2018). Blockchain Government – A Next Form of Infrastructure for the Twenty-first Century. *Journal of Open Innovation: Technology, Market, and Complexity, 4*(1), 7.

Kaur, H., & Singh, K. N. (2015). Pradhan Mantri Jan Dhan Yojana (PMJDY): A Leap Towards Financial Inclusion in India. *International Journal of Emerging Research in Management &Technology, 4*(1), 25–29.

Kaushik, A. (2017, March). GIS Based Decision Support Systems in Government: Cases from India. In *Proceedings of the 10th International Conference on Theory and Practice of Electronic Governance* (pp. 366–374). ACM.

Kumar, S. (2016). e-Governance in India. *Imperial Journal of Interdisciplinary Research (IJIR), 2*(2), 2454–1362.

Kumar, V. (2018). Social Implications of e-Government. In *Social Network Analytics for Contemporary Business Organizations* (pp. 35–50). Hershey: IGI Global.

LiveMint. (2016a). *Budget 2015: Govt Spending on Aadhaar Project to Rise 23%.* Politics. Available at http://www.livemint.com/Politics/WMU3sLfhhBousT5fLY0qRJ/Budget-2015-Govt-spending-on-Aadhaar-project-to-rise-23.html. Accessed 15 Dec 2016.

LiveMint. (2016b). *DigiLocker Partners with Road Transport and Highways Ministry.* Retrieved from http://www.livemint.com/Politics/ks0E0m7Ko-V2O78uo8FQGFK/DigiLocker-partners-with-road-transport-and-highways-ministr.html. Accessed 14 Mar 2017.

LiveMint. (2017). *Union Budget*. Retrieved from http://www.livemint.com/ Politics/HIabhPn5ApPsL8Na879uEO/Highlights-of-Union-Budget201718.html. Accessed 22 Mar 2017.

LiveMint. *22 New Schemes Under Digital India*. Retrieved from http://www.livemint.com/Home-Page/QgFspv8UzykQP99AukcSjI/Govt-launches-22-new-schemes-under-Digital-India-programme.html. Accessed 20 Mar 2017.

Martinovic, I., Kello, L., & Sluganovic, I. (2017). *Blockchains for Governmental Services: Design Principles, Applications, and Case Studies*. Centre for Technology and Global Affairs|University of Oxford.

MEITY. (2017). *Key Achievement 2017: MeitY*. Available from http://pib.nic.in/ newsite/PrintRelease.aspx?relid=174994. Accessed 20 Jan 2017.

Meity.gov.in. (2017). Retrieved from http://meity.gov.in/sites/upload_files/ dit/files/Digital%20India.pdf. Accessed 14 Mar 2017.

Ministry of Communications & Information Technology. Available on http://pib.nic.in/newsite/PrintRelease.aspx?relid=104371. Retrieved on 18 Feb 2017.

Ministry of Electronics and Information Technology. (2016). *XII Five-Year Plan on Information Technology Sector Report of Sub-Group on Cyber Security*. Cyber Laws and E-Security. Available at http://meity.gov.in/sites/upload_files/dit/ files/Plan_Report_on_Cyber_Security.pdf. Accessed 14 Nov 2016.

Misra, D., Mishra, A., Babbar, S., & Singh, S. (2017, March). Web Accessibility Assessment of Government Web Solutions: A Case Study in Digital India. In *Proceedings of the 10th International Conference on Theory and Practice of Electronic Governance* (pp. 26–34). New York: ACM.

Mitra, K. (2012). *Rise of e-Governance*. Indian Institute of Foreign Trade. Working Paper.

Mukherjee, N., & Majumder, R. (2015). Unpaid Work and Missing Women in the Indian Labour Market. *Indian Journal of Human Development, 9*(2), 191–217.

MyGov. (2016a). *Help Us Make Digital Locker Better*. Digital India. Available at https://secure.mygov.in/group-issue/help-us-make-digital-locker-better/. Accessed 15 Dec 2016.

MyGov. (2016b). *Highlights*. Government of India. Available at https://www.mygov.in/. Accessed 26 Nov 2016.

MyGov.in. https://www.mygov.in/#section1. Accessed 16 Mar 2017.

Nair, R. (2015). *Online Payment in India Accounted for 14% of Total Transaction Amount in FY 2015*. Retrieved from https://dazeinfo.com/2015/05/29/ online-payment-india-accounted-14-total-transaction-amount-fy-2015-report/. Accessed 19 Dec 2016.

Nasscom. (2016). Towards e-Governance. *Business Today*. Available at https:// www.businesstoday.in/magazine/features/it-can-be-said-that-the-role-of-ict-in-the-public-sector-has-evolved-from-informatisation-to-providing-smart-government/story/242989.html. Accessed 15 Jan 2017.

National Payments Corporation of India. (2016). *Benefits of Rupay Card*.http:// www.npci.org.in/RuPayBenefits.aspx. Accessed 27 Dec 2016.

NCOG. (2017). *National Centre for Geo Informatics.* Retrieved from https://ncog.gov.in/. Accessed 9 Mar 2017.

NeGD. (2016a). *History of NeGD.* About Us. Available at http://negd.gov.in/about-us. Accessed 15 Nov 2016.

NeGD. (2016b). *History of NeGD.* Digital India. Available at http://negd.gov.in/about-us. Accessed 10 July 2016.

NeGP. (2016). *Infrastructure.* Available at https://negp.gov.in/index.php?option=com_content&view=article&id=247&Itemid=715. Accessed 10 July 2016.

NJP, S. R., & Sahal, S. (2018). Learning Management System Under Digital India Program: Blended Learning Platform for Digital Governance. In *Marketing Initiatives for Sustainable Educational Development* (pp. 205–225). Hershey: IGI Global.

OECD. (2003). *Promise and Problems of e-Democracy, Challenges of Online Citizen Engagement.* OECD Study. Available at http://www.oecd-ilibrary.org/governance/promise-and-problems-of-e-democracy_9789264019492-en. Accessed 11 July 2016.

Palvia, S. C. J. (2013). Editorial Preface Article: e-Evolution or e-Revolution: e-Mail, e-Commerce, e-Government, e-Education. *Journal of Information Technology Case and Application Research, 15*(4), 4–12.

Palvia, S. C. J., & Sharma, S. S. (2007, December). e-Government and e-Governance: Definitions/Domain Framework and Status Around the World. In *International Conference on e-Governance* (pp. 1–12). Hyderabad, India: IECG.

Pareek, S. (2015). *12 Projects You Should Know About Under the Digital India Initiative.* The Better India. Available at http://www.thebetterindia.com/27331/12-projects-you-should-know-about-under-the-digital-india-initiative/. Accessed 10 July 2016.

Pathak, D. C., & Mishra, S. (2015). Poverty Estimates in India: Old and New Methods, 2004–2005. *Poverty & Public Policy, 7*(1), 44–63.

Paul, Y. (2007). Role of Genetic Factors in Polio Eradication: New Challenge for Policy Makers. *Vaccine, 25*(50), 8365–8371.

PESA. (2017). Retrieved from http://pesadarpan.gov.in/home. Accessed 9 Mar 2017.

Planning Commission India. (2017). Available at http://wwww.planningcommission.gov.in. Accessed 26 Aug 2017.

PMINDIA. (2016). *Mann Ki Baat.* Available at http://www.pmindia.gov.in/en/tag/mann-ki-baat/. Accessed 26 Nov 2016.

Poulose, V. A. (2010). e-Governance and Infrastructure: Looking Ahead. *RITES Journal, 1*(3), 9–1.

Prabhakaran, D., Jeemon, P., Sharma, M., Roth, G. A., Johnson, C., Harikrishnan, S., et al. (2018). The Changing Patterns of Cardiovascular Diseases and Their Risk Factors in the States of India: The Global Burden of Disease Study 1990–2016. *The Lancet Global Health, 6*(12), e1339–e1351.

Pradhan Mantra Jan-Dhan Yojana. (2016). *Progress-Report*. Retrieved from http://pmjdy.gov.in/account. Accessed 27 Dec 2016.

Prasad, R., & Gupta, R. (2017). *e-Commerce and e-Consumer in India*. Report for Federation of Indian Electronic Commerce & Industry & Ministry of Consumer Affairs, Government of India.

Press Information Bureau. (2016). *e-Governance Vision 2020 to Provide Inclusive, Integrated Single Window View of Services to All Stakeholders: Dr. Nasim Zaidi*. Election Commission. Available at http://pib.nic.in/newsite/PrintRelease. aspx?relid=136655. Accessed 15 Nov 2016.

Punj, S. (2012). *A Number of Changes*. Business Today. Available at http://www. businesstoday.in/magazine/features/uid-project-nandan-nilekani-future-unique-identification/story/22288.html. Accessed 15 Dec 2016.

Purohit, B. C. (2012). Budgetary Expenditure on Health and Human Development in India. *International Journal of Population Research, 2012*, 1–13.

Rahman, H., & Ramos, I. (2014). Stages of e-Government Maturity Models: Emergence of e-Governance at the Grass Roots. In *Technology Development and Platform Enhancements for Successful Global e-Government Design* (pp. 228–246). Hershey: IGI Global.

Rajadhyaksha, A. (2013). *In the Wake of Aadhaar the Digital Ecosystem of Governance in India*. Centre for the Study of Culture & Society. Available at http://www.jnu.ac.in/SSS/CSSP/RaviShukla.pdf. Accessed 15 Dec 2016.

Ramachandran, N. (2011). Are Women Equally Unequal in India? Looking Across Geographic Space. *Gender, Technology and Development, 15*(3), 363–387.

Rana, N. P., Dwivedi, Y. K., Williams, M. D., & Weerakkody, V. (2016). Adoption of Online Public Grievance Redressal System in India: Toward Developing a Unified View. *Computers in Human Behavior, 59*, 265–282.

Rao, V. R. (2018). Public Services Through Multi-Channel Issues and Challenges: A Case From India. In *Optimizing Current Practices in e-Services and Mobile Applications* (pp. 273–300). Hershey: IGI Global.

Rao, G. K., & Dey, S. (2011). Text Mining Based Decision Support System (TMbDSS) for e-Governance: A Roadmap for India. In *Advances in Computing and Information Technology* (pp. 270–281). Berlin/Heidelberg: Springer.

Retrieved from http://www.hindustantimes.com/mumbai-news/cbse-starts-online-repository-for-students-to-view-mark-sheets/story-YZxvP2x3sk-PjRrDcK5rMeO.html. Accessed 15 Mar 2017.

Sabharwal, M., & Berman, E. M. (Eds.). (2013). *Public Administration in South Asia: India, Bangladesh, and Pakistan*. Boca Raton: CRC Press.

Sapru, R. K., & Sapru, Y. (2014). Good Governance Through e-Governance with Special Reference to India. *Indian Journal of Public Administration, 60*(2), 313–331.

Sharma, M., & Gupta, R. (2015). Basis Function and PN Phase Generation in TDCS and WDCS Towards Dynamic Spectrum Access. In *2015 Fifth*

International Conference on Communication Systems and Network Technologies (pp. 364–368). Gwalior, India: IEEE Xplore.

Shrivastava, S., & Pal, S. N. (2017, March). A Big Data Analytics Framework for Enterprise Service Ecosystems in an e-Governance Scenario. In *Proceedings of the 10th International Conference on Theory and Practice of Electronic Governance* (pp. 5–11). New York: ACM.

Singh, J. P. (2012). *e-Governance in India: Existing Context and Possible Scope for UNDP Programing Over 2013–18.* IT for Change. Available at http://www.itforchange.net/E-governance_in_India%3A_Existing_context_and_possible_scope_for_UNDP_programing_over_2013-18. Accessed 10 July 2016.

Sood, T. (2017). Services Marketing: A Sector of the Current Millennium. In *Strategic Marketing Management and Tactics in the Service Industry* (pp. 15–42). Hershey: IGI Global.

Srivastava, A. K. (2014). Applying Service Quality Metrics for e–Urban Governance: A Case Study of Lucknow Municipal Corporation. In *Governometrics and Technological Innovation for Public Policy Design and Precision* (pp. 232–266). Hershey: IGI Global.

Swathi, P. (2016). *Role of IT Infrastructure in Public Service Delivery.* Centre for Innovations in Public Systems. Available at http://www.cips.org.in/documents/CSCs_WS/Kadapa/ITInfrastructure.pdf. Accessed 10 July 2016.

TCS. (2008). *TCS Moots Increased Public Spending on Transparency in Governance.* Available at https://www.domain-b.com/companies/companies_t/tcs/20080610_public_spending.html. Accessed 15 Dec 2016.

The Asset Directory. (2017). Retrieved from http://assetdirectory.gov.in/. Accessed 9 Mar 2017.

The Business Standards. (2017). BHIM App. Retrieved from http://www.business-standard.com/article/economy-policy/bhim-app-breaks-all-records-crosses-18-million-downloads-117031700305_1.html. Accessed 17 Mar 2017.

The Economic Times. (2014). *Attendance.gov.in: Modi Government Launches Website to Track Attendance of Government Employees.* Retrieved from http://economictimes.indiatimes.com/news/politics-and-nation/attendance-gov-in-modi-government-launches-website-to-track-attendance-of-government-employees/articleshow/44588559.cms. Accessed 16 Mar 2017.

The Economic Times. (2017). Retrieved from http://economictimes.indiatimes.com/industry/tech/software/hp-launches-centre-of-excellence-to-support-digital-india/articleshow/57533741.cms. Accessed 10 Mar 2017.

The Financial Express. (2014). *eBharath 2020: A Vision for Future e-Governance.* Corporates & Markets. Available at http://www.financialexpress.com/archive/ebharath-2020-a-vision-for-future-e-governance/1254379/. Accessed 15 Nov 2016.

The Financial Express. (2017). *'Digital India': Top 10 initiatives announced by Centre.* Available at http://www.financialexpress.com/economy/digital-india-top-10-initiatives-announced-by-centre/185308/. Accessed 9 Mar 2017.

The Hindu. (2014). *Modi Goes on AIR*. National. Available at http://www.the-hindu.com/news/national/modis-first-radio-interaction-mann-ki-baat-on-all-india-radio/article6468709.ece. Accessed 26 Nov 2016.

The Ministry of Electronics and Information Technology. (2017). Retrieved from https://apps.mgov.gov.in/index.jsp. Accessed 01 Feb 2017.

The Statista Portal. (2017). *Number of Mobile Users*. Retrieved from https://www.statista.com/statistics/274658/forecast-of-mobile-phone-users-in-india/. Accessed 01 Feb 2017.

The Times of India. (2016). '*PM Modi to Address 16th Edition of Mann Ki Baat on Sunday*. India. Available at http://timesofindia.indiatimes.com/india/PM-Modi-to-address-16th-edition-of-Mann-Ki-Baat-on-Sunday/articleshow/50726987.cms. Accessed 26 Nov 2016.

The World Bank. (2016). *GDP Per Capita (Current US$)*. Retrieved from http://data.worldbank.org/indicator/NY.GDP.PCAP.CD?end=2015&locations=IN-XM&start=1960&view=chart. Accessed 19 Dec 2016.

Times of India. (2017). *Mann Ki Baat. Key Performance Indicators*. Retrieved from http://timesofindia.indiatimes.com/india/mann-ki-baat-india-scripts-world-record-pm-narendra-modi-congratulates-isro/articleshow/57354562.cms. Accessed 22 Mar 2017.

Toshio, O. B. I. (2015). *2015 Waseda – IAC International e-Government Ranking Survey*. Institute of e-Government, Waseda University. Available at http://e-gov.waseda.ac.jp/pdf/2015_Waseda_IAC_E-Government_Press_Release.pdf. Accessed 16 July 2016.

UN.Org. (2017). *UN e-Government Survey 2016*. Retrieved from https://public-administration.un.org/egovkb/en-us/Reports/UN-E-Government-Survey-2016. Accessed 22 Mar 2017.

UNESCO. (2017). *What Is e-Governance?* Retrieved from http://portal.unesco.org/ci/en/ev.phpURL_ID=4404&URL_DO=DO_TOPIC&URL_SECTION=201.html. Accessed 22 Mar 2017.

United Nations. (2005). *United Nations Global e-Government Readiness Report 2005: From e-Government to e-Inclusion*. United Nations Department of Economic and Social Affairs. Available at https://publicadministration.un.org/egovkb/Portals/egovkb/Documents/un/2005-Survey/Complete-survey.pdf. Accessed 16 July 2016.

United Nations. (2008). *United Nations e-Government Survey 2010-From e-Governance to Connected Governance*. United Nations Department of Economic and Social Affairs. Available at https://publicadministration.un.org/egovkb/portals/egovkb/documents/un/2008-survey/unpan028607.pdf. Accessed 16 July 2016.

United Nations. (2010). *United Nations e-Government Survey 2010 Leveraging e-Government at a Time of Financial and Economic Crisis*. United Nations Department of Economic and Social Affairs. Available at https://publicadministration.un.org/egovkb/portals/egovkb/documents/un/2010-survey/complete-survey.pdf. Accessed 16 July 2016.

United Nations. (2012). *United Nations e-Government for the People e-Government Survey 2012*. United Nations Department of Economic and Social Affairs. Available at https://publicadministration.un.org/egovkb/Portals/egovkb/Documents/un/2012-Survey/Complete-Survey.pdf. Accessed 16 July 2016.

United Nations. (2014a). *United Nations e-Government Survey 2014- e-Government for the Future We Want*. United Nations Department of Economic and Social Affairs. Available at https://publicadministration.un.org/egovkb/portals/egovkb/documents/un/2014-survey/e-gov_complete_survey-2014.pdf. Accessed 16 July 2016.

United Nations. (2014b). *United Nations e-Government Survey 2014*. United Nations Department of Economic and Social Affairs. Available at https://publicadministration.un.org/egovkb/portals/egovkb/documents/un/2014-survey/e-gov_complete_survey-2014.pdf. Accessed 11 July 2016.

United Nations. (2016). *What Is Good Governance*. Available at http://www.unescap.org/sites/default/files/good-governance.pdf. Accessed 11 July 2016.

United Nations Human Rights. (2016). *Good Governance and Human Rights*. Good Governance, Development. Available at http://www.ohchr.org/EN/Issues/Development/GoodGovernance/Pages/GoodGovernanceIndex.aspx. Accessed 11 July 2016.

UNPAN. (2003). *e-Government Strategy*. Implementing the President's Management Agenda for e-Government. Available at http://unpan1.un.org/intradoc/groups/public/documents/other/unpan032102.pdf. Accessed 11 July 2016.

Verma, N., & Mishra, A. (2009, November). india.gov.in-India's Approach in Constructing One-stop-solution Towards e-Government. In *Proceedings of the 3rd International Conference on Theory and Practice of Electronic Governance* (pp. 247–252). New York: ACM.

Verma, N., Singh, S., & Misra, D. P. (2007, December). Citizen Participation in the Process of ICT Enabled Governance: A Case Study. In *Proceedings of the 1st International Conference on Theory and Practice of Electronic Governance* (pp. 371–379). New York: ACM.

Yadav, N., & Singh, V. B. (2013). e-Governance: Past, Present and Future in India. *International Journal of Computer Applications, 53*(7), 36–48.

Yerramilli, R., & Swamy, N. K. (2017, August). Mobile Governance—A Complement for Successful e-Governance (Study on Challenges in Mobile Governance). In *2017 International Conference on Smart Technologies for Smart Nation (SmartTechCon)* (pp. 1549–1554). Bangalore, India: IEEE.

INDEX

© The Author(s) 2019
S. K. Muttoo et al., *E-Governance in India*,
https://doi.org/10.1007/978-981-13-8852-1

GPSR Compliance
The European Union's (EU) General Product Safety Regulation (GPSR) is a set
of rules that requires consumer products to be safe and our obligations to
ensure this.

If you have any concerns about our products, you can contact us on

ProductSafety@springernature.com

In case Publisher is established outside the EU, the EU authorized
representative is

Springer Nature Customer Service Center GmbH
Europaplatz 3
69115 Heidelberg, Germany